Protein purification applications

a practical approach

TITLES PUBLISHED IN
THE
PRACTICAL APPROACH
SERIES

Series editors:
Dr D Rickwood
Department of Biology, University of Essex
Wivenhoe Park, Colchester, Essex CO4 3SQ, UK
Dr B D Hames
Department of Biochemistry, University of Leeds
Leeds LS2 9JT, UK

Affinity chromatography
Animal cell culture
Antibodies I & II
Biochemical toxicology
Biological membranes
Carbohydrate analysis
Cell growth and division
Centrifugation (2nd Edition)
Computers in microbiology
DNA cloning I, II & III
Drosophila
Electron microscopy
in molecular biology
Fermentation
Gel electrophoresis
of nucleic acids (2nd edition)
Gel electrophoresis of proteins
Genome analysis
HPLC of small molecules
HPLC of macromolecules
Human cytogenetics
Human genetic diseases
Immobilised cells and enzymes
Iodinated density gradient media
Light microscopy in biology
Liposomes
Lymphocytes
Lymphokines and interferons
Mammalian development
Medical bacteriology
Medical mycology

Microcomputers in biology
Microcomputers in physiology
Mitochondria
Mutagenicity testing
Neurochemistry
Nucleic acid and
protein sequence analysis
Nucleic acid hybridisation
Nucleic acids sequencing
Oligonucleotide synthesis
Photosynthesis:
energy transduction
Plant cell culture
Plant molecular biology
Plasmids
Prostaglandins
and related substances
Protein function
Protein purification applications
Protein purification methods
Protein sequencing
Protein structure
Proteolytic enzymes
Ribosomes and protein synthesis
Solid phase peptide synthesis
Spectrophotometry
and spectrofluorimetry
Steroid hormones
Teratocarcinomas
and embryonic stem cells
Transcription and translation
Virology
Yeast

Protein purification applications

a practical approach

Edited by

E L V Harris and S Angal

Celltech Ltd, 216 Bath Road,
Slough SL1 4EN, UK

IRL PRESS
—at—
OXFORD UNIVERSITY PRESS
Oxford New York Tokyo

Oxford University Press
Walton St., Oxford OX2 6DP

© Oxford University Press 1990

First published 1990

British Library Cataloguing in Publication Data

Protein purification applications
  1. Proteins. Purifications
  I. Harris,E.L.V. II. Angal,S. III. Series
  547.7'5

Library of Congress Cataloging in Publication Data

Protein purification applications: a practical approach / edited by
  E.L.V. Harris and S. Angal.
    (Practical approach series)
    Companion v. to: Protein purification methods / edited by E.L.V. Harris
  and S. Angal. 1989.
    Includes bibliographies and index.
    1. Proteins—purification. I. Harris, E. L. V. II. Angal, S.
  III. Protein purification methods. IV. Series.
    [DNLM: 1. Proteins—isolation & purification. QU 55 P968065]
  QP551.P697514 1989
  574. 19'296—dc20   89-11109

ISBN 0 19 963022 4
ISBN 0 19 963023 2 (Pbk)

Previously announced as:
ISBN 1 85221 163 6
ISBN 1 85221 162 8 (Pbk)

Typeset and printed by Information Press Ltd, Oxford, England.

Preface

Protein purification is central to many areas of biochemical research. With the recent advances in gene cloning and expression it is not only protein biochemists who need to be able to purify proteins. Despite its key role protein purification is not a subject studied in depth in many undergraduate courses. Our experience of training graduates and post-graduates has led to the realization of the often limited knowledge and experience in this field. So when the subject of editing a 'Practical Approach' book on protein purification came up there was an immediate attraction. As we began the detailed planning of the book it became apparent that to cover this broad field adequately would entail division of the contents into two volumes, one on basic methods and one on applications. We have aimed to cover both the fundamentals and the more recent advances in the field. We therefore anticipate that both volumes will be useful as bench manuals to a wide range of biochemists and molecular biologists, from novices to experienced protein purifiers. In this the second volume, *Protein purification applications,* we have covered several specific applications of protein purification. This builds on the techniques described in the first volume and demonstrates how they can be combined to achieve a variety of objectives.

The complexities and problems of scaling-up a purification process are frequently not understood. The need of the biotechnology industry to solve many of these problems has driven advances in equipment and matrix design. Chapter 1 discusses the considerations to be taken into account when scaling-up a process. In Chapter 2 purification of proteins for therapeutic use is covered. This application places stringent requirements on the purity of the protein, and therefore restricts the types of purification methods which can be used and also requires very sensitive methods for the assay of potential contaminants. Basic techniques for protein crystallization are described in Chapter 3, together with a discussion of the influence of protein heterogeneity on crystal structure, a subject rarely covered in textbooks but becoming more important as the number and purity of proteins available to scientists increases. Purification of membrane proteins often necessitates adaptation of the traditional protein purification methods. The techniques required are discussed in Chapter 4 and also in another book of this series (*'Biological membranes: a practical approach'* Edited by J.B.C. Findlay and W.H. Evans). Protein sequencing and the constraints it places on purification are discussed briefly in Chapter 5, and are also the subject of another book in this series (*'Protein sequencing: a practical approach'* Edited by J.B.C. Findlay and M.J. Geisow). Once some sequence analysis has been obtained for a protein the gene can be cloned and expressed, allowing purification of larger quantities of the protein. During cloning the gene can be manipulated to aid purification by producing a fusion protein as discussed in Chapter 6. In the final chapter several examples of purification processes are covered. The proteins range from secreted to intracellular, and from the abundant to the less abundant. These examples are used, not only to allow the reader to purify the specific proteins included, but also to demonstrate by example how typical purification processes are designed. This latter topic is frequently omitted from other texts on protein purification; we have therefore covered it in both volumes.

We are indebted to Margaret Turner for her expert secretarial assistance without which the book would not have reached publication. We are also grateful to Martin and Shantanu for their moral support over the two years it has taken to complete the book.

E.L.V.Harris
S.Angal

Contributors

S.Angal
Celltech Ltd, 216 Bath Road, Slough SL1 4EN, UK

J.A.Asenjo
Department of Biochemical Engineering, University of Reading, P.O. Box 226, Reading RG6 2AP, UK

S.J.Brewer
Mail Zone GG4B, Monsanto Company, 700 Chesterfield Village Parkway, St. Louis, MO 63198, USA

A.F.Bristow
National Institute of Biological Standards and Control, Blanche Lane, South Mimms, Potters Bar, Herts EN6 3QG, UK

V.C.Duance
AFRC Institute of Food Research, Langford, Bristol BS18 7DY, UK

J.Fedor
Chiron Corporation, 4560 Horton Street, Emeryville, CA 94608, USA

J.B.C.Findlay
Department of Biochemistry, University of Leeds, Leeds LD2 9JT, UK

S.J.Froud
Celltech Ltd, 216 Bath Road, Slough SL1 4EN, UK

C.George-Nascimento
Chiron Corporation, 4560 Horton Street, Emeryville, CA 94608, USA

E.L.V.Harris
Celltech Ltd, 216 Bath Road, Slough SL1 4EN, UK

H.Kirby
Celltech Ltd, 216 Bath Road, Slough SL1 4EN, UK

G.Murphy
Strangeways Research Laboratory, Worts' Causeway, Cambridge CB1 4RN, UK

I.Patrick
Celltech Ltd, 228 Bath Road, Slough SL1 4EN, UK

M.Perry
Celltech Ltd, 240 Bath Road, Slough SL1 4EN, UK

H.M.Sassenfeld
Immunex Co., 51 University Street, Seattle, WA 98101, USA

S.P.Wood
Department of Crystallography, Birkbeck College, University of London, Malet Street, London WC1E 7MX

Contents

Abbreviations

ATP	adenosine triphosphate
CDR	cell debris remover
CM	carboxymethyl
CMC	critical micellar concentration
CPB	carboxypeptidase B
CTAB	cetyl trimethylammonium bromide
DEAE	diethylaminoethyl
DNA	deoxyribonucleic acid
DOC	deoxycholate
DTAB	dodecyl trimethylammonium bromide
DTT	dithiothreitol
EDTA	ethylenediamine tetraacetate
EGF	epidermal growth factor
FAB-MS	fast atom bombardment mass spectrometry
FACE	formic acid−acetic acid−chloroform−ethanol (1:1:2:1 v/v)
FE	formic acid−ethanol
FPLC	fast protein liquid chromatography
Hepes	N,2-hydroxyethylpiperazine-N'-ethane sulphonic acid
HPLC	high performance liquid chromatography
LAL	*Limulus* amboecyte lysate
LDL	low density lipoprotein
LPS	lipopolysaccharide
MAP	mouse antibody production test
Mes	2-(N-morpholino)ethane sulphonic acid
Mops	morpholinopropane sulphonic acid
NAD	nicotinamide adenine dinucleotide
NADP	nicotinamide adenine dinucleotide phosphate
NMR	nuclear magnetic resonance
NSB	non-specific binding
PAP	peroxidase−anti-peroxidase
PBS	phosphate buffered saline
PDGF	platelet derived growth factor
PEG	polyethylene glycol
pI	isolectric point
p.p.m.	parts per million
PTFE	polytetrafluoroethylene
RIA	radioimmunoassay
SDS	sodium dodecyl sulphate
SDS-PAGE	sodium dodecyl sulphate polyacrylamide gel electrophoresis
TGF(α or β)	transforming growth factor (α or β)
TMPD	N,N,N',N'-tetramethyl-p-phenylenediamine
TNF	tumour necrosis factor
Tris	tris(hydroxymethyl)methylamine

CHAPTER 1

Large-scale protein purification

J.A.ASENJO and I.PATRICK

1. INTRODUCTION

Any description of the problems of scaling up a protein purification process is complicated by the difficulty that 'large-scale' represents a moving target. The modern
biotechnology industry produces specialist proteins and peptides (e.g. monoclonal antibodies and rDNA polypeptides) in batches of grams to kilograms and utilizes
chromatographic columns ranging from litres to thousands of litres. Obviously, it is
difficult to define where small-scale ends and 'large-scale' begins.

   Purification processes are generally designed by laboratory scientists. This chapter
is intended for scientists who are faced with transferring laboratory processes to the
industrial scale. Scaling up is in the realm of engineering and hence involves many
considerations which are not associated with pure science as carried out in the laboratory.
The laboratory protein purification often involves learning and finding out facts about
a problem in a highly defined situation—certainly a situation in which experiments can
be repeated. The move to plant scale allows only one experiment which must work
and, on the other hand, for each question there is a large number of alternative,
apparently acceptable, answers. The choice of solution to a given problem thus involves
a complex balancing act which requires judgement and experience. It is not the purpose
of this chapter to review the discipline of protein purification which is addressed in
the companion volume, *Protein Purification Methods: A Practical Approach* (1). The
purpose is, firstly, to help the scientist/engineer to recognize which questions need to
be answered and to gain some insight into how such answers might be obtained. This
chapter shows some of the principles to be followed to design a preliminary protein
purification strategy; it describes the use and applicability of large-scale operations (scale-
up); as well as the state of the art on optimization of protein purification processes
including the use of modern computer techniques; it also gives some information on
economics and costs. Finally we intend to draw attention to some pitfalls encountered
on the large scale which might not be obvious from the laboratory bench.

2. PROCESS PHILOSOPHY

At the point at which problems regarding the scale of operation are initially addressed,
usually many key decisions have already been taken. The target protein, the source
and production method, the required level of purity and some estimate of the eventual
quantity of finished product required will have been considered. Significantly, the
purification process is the stage most likely to require radical re-evaluation and change.

   Practically any project involving purification of a protein will already have been subject

1

to some degree of scale-up, both in quantity and quality, while at the laboratory bench. A few micrograms of crude extract to test for an *in vitro* biological response may be scaled up to a few milligrams of essentially pure product, perhaps for physico-chemical characterization or animal testing. Unless the real problems of scale-up have been addressed by this stage the project is likely to run into major problems later. The basic fault with the general laboratory approach is that it may overlook the fact that the aims and philosophy of the large- and small-scale operations are very different.

The aim of the small-scale operation is to produce material adequate to prove the project feasible in some limited way. The process is a 'hands-on' one, with close and informed monitoring of each stage (over and above that required exclusively for process control) with individual attention from highly skilled and motivated personnel at all times. Under these circumstances, processes subject to very precise and critical control of various parameters can be carried out successfully. Where necessary, *ad hoc* recycling is not too great a problem. As the demand for the product grows, the process moves to the larger scale as a matter of course.

On the large scale, an essential aim is to maximize yield and minimize cost. Human intervention must be minimized and preferably restricted to trained operators, not qualified and experienced scientists. Not only must the process be feasible on the large scale, but variations in the original feedstock and process within experimental error must be absorbed by the subsequent stages. Production plants are run to a predetermined schedule and, apart from regulatory constraints for therapeutic products, the need to reprocess 'out-of-specification' material is extremely cumbersome and should be avoided.

In addition to being both robust and effective, the process needs to be as abbreviated as possible. The initial purification process is inevitably devised at a stage when knowledge and experience of the protein and the particular purification problems are at a minimum. Specifications are either not clear or not known; the adequacy and relevance of various forms of assay (bioassay, SDS-PAGE, UV, HPLC) are not understood. Inevitably this, combined with the step-by-step process development often adopted, leads to over-elaborate processes where one or more stages may be quite unnecessary. Unnecessary stages are not only uneconomic, but problems are perpetuated and multiplied. This defeats the basic aims of scaling up.

Finally, the long-term goal of the whole project may mean that work on the laboratory scale has implications which are not immediately obvious to the scientist involved. The modern biotechnology industry is extremely competitive and time constraints are critically important. Proteins and polypeptides manufactured and purified on a very modest scale may be tested in animal studies and then in restricted clinical trials over a number of months. The results are then presented to regulatory authorities to support and justify further large-scale clinical trials. Any but the most minor of revisions to the process after this stage may disqualify the preliminary clinical data. If the process is not already properly designed for large-scale operation, the project manager is faced with a difficult choice: either scale up an uneconomic and over-elaborate process with all the problems that this implies, or start the project again to the substantial advantage of competitors. It is clearly of critical importance to review any small-scale process at a very early stage with a view to employing steps amenable to scale-up and introducing these at the laboratory bench.

3. PROCESS SELECTION

The first stage, then, is to define realistic separation steps. We can describe our target process immediately as a multistage chromatographic process with intermediate conditioning stages, such as concentration, pH changes and ionic strength changes. Selection of this process is achieved by using all the information available on the target protein and background impurity matrix and by following five main rules.

(i)   Choose separation processes based on different properties.
(ii)   Choose conditions that will exploit the greatest differences in those properties between the product and the background impurity matrix.
(iii)   Separate the most plentiful impurities first.
(iv)   Use a highly selective step as soon as possible.
(v)   Carry out the most arduous or expensive step last.

   This approach has been used successfully in chemical process synthesis (2). Some of the heuristics used for protein purification are based on those developed in chemical engineering for separation processes. An adaptation of these has been recently proposed for biotechnological separations (3). Here we have summarized the ones we believe to be the most important. Like all rules of thumb, they need to be applied as guidelines with a generous measure of common sense.

(i) *Rule 1*. It is to be expected that in a sequence of two or three separation steps one should obtain a higher overall purification if each of these operations is based on a different physicochemical property of the material to be separated: for example, ion-exchange followed by affinity chromatography or affinity chromatography followed by gel-filtration. On the other hand, it is common to use an ion-exchange column as a preliminary concentration and crude purification step followed by a more selective high-resolution ion-exchange step.

(ii) *Rule 2*. It is quite evident that application of this rule is likely to make separation easier and therefore more successful. It presupposes to some extent, however, that there is background knowledge of the general physicochemical properties not only of the product but also of the main impurities present, which is not always the case.

(iii) *Rule 3*. Separation of the most plentiful impurities at the earliest possible stage should lead to the greatest concentration in terms of mass or volume. Since the main component of bioprocess streams is likely to be water, this stresses that an early step in the separation process should be concentration of the protein and removal of bulk water. Not only does this make the product stream much more convenient to handle, but it will also lead to a proportional reduction in the time required for subsequent steps, chromatographic or otherwise. Furthermore, in the process of removing this major impurity, many minor contaminants will be removed simultaneously—ultrafiltration removes other small molecules in addition to water; ion exchange chromatography may separate from the product not only a major protein impurity such as BSA, but also other minor proteins and a large proportion of nucleic acid contamination inevitably present in crude extracts or fermentation liquors.

(iv) *Rule 4*. This is a crucial rule and one most easily overlooked. Protein purification processes commonly have many steps, and more than are necessary. The early use of a high-resolution step may dramatically reduce the total number of subsequent stages, making these into 'polishing' operations. Firstly, it should be remembered that the intention is to aim for a highly selective operation to separate the product from the background matrix rather than to accomplish high-resolution separation of the contaminant proteins into well-defined but valueless fractions. Secondly, we should observe that early users of highly selective affinity columns (for example monoclonal-antibody) often employed these as essentially the last stage of purification rather than the reverse. The justification for this seems to have been that the columns were very expensive and needed protection from irreversible contamination of one sort or another. On the small scale or in an academic environment this seems to make good sense, but on the large production scale where overheads are an extremely important factor of the economic calculation, the opposite view should be considered.

(v) *Rule 5.* When process volumes are large, it is necessary to ensure that the processes employed are physically appropriate. Adsorption−desorption processes such as affinity chromatography rather than gradient elution from ion exchange columns are much more suitable for bulk processing. Similarly, it would require very compelling reasons indeed to justify the use of an analytical gel filtration chromatography column very early in the purification process rather than at the very last stage as this is an inefficient technique for protein separation. On the other hand, a gel filtration desalting procedure might be appropriate at either stage.

There are usually not more than four or five necessary main steps in a large-scale protein purification procedure.

(i)   Cell separation;
(ii)   cell disruption and debris separation (for intracellular proteins only);
(iii)   concentration;
(iv)   pretreatment or primary isolation;
(v)   high-resolution purification;
(vi)   polishing of final product.

It is important to bear in mind that there will be interactions between the fermentation system and protein recovery, hence production methods used will affect the later purifi-cation steps (4−6). Important variables in production will be the type of fermentation system used (bacterial, yeast or mammalian), the medium composition (presence of calf serum or use of serum-free media, presence of protease inducers) and the type of reactor used, for example a membrane reactor for cell retention or immobilized product fermentation (4). It is important then, when choosing the fermentation system, fermentation conditions and reactor system, to carefully evaluate the effect that these decisions will have on the downstream processing and protein purification stages. Product concentration will partly depend on the reactor system used. Presence of proteases, as well as bacterial contamination, have to be minimized, which results in the need for rapid processing requirements. Presence of calf serum will increase the number of purification stages required. Recombinant proteins expressed in *Escherichia coli* are in many cases present in particles that need to be solubilized and renatured (7). In con-clusion, not only is it important to optimize fermentation yield *per se,* but also to present the product in a form which allows optimum overall yield.

The problems that have to be solved in process selection and optimization of downstream operations thus fall into two categories: (i) choice of alternative operations, for example between gel filtration and diafiltration; and (ii) the design of an optimal (for example chromatographic) sequence with maximal yield and minimal number of steps. These problems are generally addressed by empirical methods based on experience but recently computer-based expert systems have been proposed (8; see Section 10 of this chapter).

4. PROTEIN HANDLING

A number of problems are specific to protein solutions in addition to the normal bulk handling and transfer difficulties of large volumes of liquid. They can be summarized as follows:

(i)   denaturation and surface effects;
(ii)   concentration and aggregation;
(iii)   stability and degradation;
(iv)   cleaning and sanitation.

Although these are known and recognized in any laboratory working with proteins on the small scale, they take new and problematic forms on the large scale.

4.1 Denaturation and surface effects

Denaturation may be due to a variety of causes, including temperature, extremes of pH and surface effects. The first two of these are quite well known phenomena. The latter can be frequently overlooked or ignored at the smaller scale. It may become a major problem when selecting pumps for the large scale when moving up from the laboratory where peristaltic pumps are probably most commonly used.

In general, proteins have both hydrophobic and hydrophilic regions and are surface-active, tending to spread out and denature at accessible interfaces. When the surface is static and small in relation to the bulk solution, this effect is scarcely perceptible. On air−water interfaces, and in circumstances where gas is finely divided with a large surface area or in violent motion, as in the vortex of a centrifugal pump, this denaturation effect can become very serious. The violent agitation causes removal of protein from the gas−water interface providing fresh surface for more protein. The previously adsorbed protein may be partially denatured and ultimately revealed as aggregates. It may also be visible as a finely divided precipitate eventually leading to blocked in-line filters and adding to overall problems and losses. In extreme cases, practically all the protein in solution may be destroyed in this way.

4.2 Concentration and aggregation

Many proteins in solution tend to aggregate into dimers or higher oligomers by a variety of mechanisms. The effect is more pronounced at higher concentrations, and as a result of this, on the small scale, the problem may be controlled by maintaining operations at low concentrations; $1-3$ mg ml^{-1} is typical.

On the large scale this represents up to 1000 litres per kg protein, which begins to represent appreciable problems in terms of the time required to transfer solutions through

Table 1. Increase in preferred working concentration with scale-up.

Throughput (g per cycle)	Concentration (mg ml^{-1})
1−100	c. 1−3
200−500	c. 5−10
1000−5000	c. 20+

equipment, particularly through chromatographic systems, and in the economics of large-scale plant, in terms of both capital and overhead.

Therefore in practice there is a steady increase in preferred working concentration with scale-up (*Table 1*). However, the higher concentrations lead to higher aggregate levels and higher viscosities, and this in turn to deterioration in chromatographic performance. At the least, the aggregates represent a nuisance and a small loss of product. At the worst, they lead to the need for a final gel filtration chromatography step which will increase the cost of the process and will also decrease the overall yield.

4.3 Stability and degradation

Protein solutions degrade in a variety of ways. These include loss of N-terminal amino acids due to exopeptidases, cleavage of peptide bonds by endopeptidases and damage to side chains by loss of carbohydrate substituents and amide nitrogen.

Control of these problems may be addressed by employing low temperatures (e.g. 4°C) at which enzymic and microbiological activity is minimized, low concentrations and short processing times. Clearly these are contradictory to some extent and may be difficult to balance on economic grounds. In any case, actual microbiological contamination should be controlled at all times by both careful cleaning and sanitation and by periodic microporous filtration. Prior knowledge of the stability of the protein with temperature, pH, ionic strength and proteolysis will give important information when designing the purification process. Thus, maximum limits on processing times and conditions will be defined. On a small scale, short processing times are easily achieved; on scale-up, processing times may well increase significantly.

4.4 Cleaning and sanitation

It is appropriate to consider cleaning and sanitation with general protein handling problems. All process equipment needs periodic cleaning (removal of dirt) and sanitation (removal of 'bioburden'). Exactly how clean, or how low, the 'bioburden' needs to be depends to some extent on the ultimate use of the product, but even a low level of microbiological contamination of process streams can be at the very least a great nuisance in terms of reduced life of filters and chromatographic packings and reduced product stability. Some attention must be paid to this problem at the development stage in selecting cleaning agents and routines compatible with equipment materials of construction, chromatographic packing materials, ultrafilters and product safety.

5. PROTEIN SOLUTION TRANSFER

Product transfer on the large scale implies the use of pumps and piping to transfer the product solution around a system. On the laboratory scale transfers are very often made by manually moving beakers from one location to another and by peristaltic, or less commonly, positive displacement pumps. On an industrial scale, pumps are available in a wide variety of both size and operational principle. For a particular stage in the process, pumping requirements should be specified according to criteria such as:

(i)   *performance:* pressure, flow rate, constancy of flow;
(ii)   *materials of construction:* wetted parts;
(iii)   *external finish*: electrical protection, flameproofing, corrosion protection.

Most of these requirements can be readily deduced from the laboratory scale, others can be determined empirically and based on experience. Additional criteria must be applied on the larger scale where making the correct choice the first time is critical. This must include consideration of both air and cavitation effects, together with cleaning and sanitation.

There is a widespread conception that proteins in solution are very susceptible to denaturation by shear effects while pumping, and that less damaging pumps (such as peristaltic ones) owe their effectiveness to their low shear. In fact, proteins are much more resistant to shear than generally believed and the greater hazard is posed by entrapped air. This has already been discussed earlier in this chapter.

In view of their low cost and ready availability in a variety of materials and convenience, centrifugal and impeller pumps are appropriate for buffer and exhaust streams, but product streams should always be handled by positive displacement pumps such as rotary lobe, rotary piston or progressive cavity pumps.

As mentioned earlier, the problem with centrifugal pumps is that they tend to entrap air in the central vortex of the pump chamber. It is this violently agitated vortex that tends to denature proteins passing through the equipment. Impeller pumps have essentially similar problems. Lobe pumps, on the other hand, are designed with two interlocking rotors turning inside a cavity, the shape of which approximates to a figure eight. The inlet and outlet ports are situated on opposite sides of the narrow part of the eight. As each rotor, which might have 2−4 lobes, rotates so the advancing lobes pass around the outer part of the cavity, scooping liquid forwards. The returning lobe interlocks with a hollow on the neighbouring rotor and therefore does not move fluid back. Thus the pump takes large 'bites' of liquid in a very gentle manner with minimal mechanical action.

Piping and valves may be selected on the same preliminary criteria as pumps. Larger pilot-scale equipment is available with materials of construction identical with those used in the laboratory, namely food-grade polypropylene and PTFE piping and joint systems. Although these are enormously convenient in terms of ease of assembly and disassembly, PTFE systems in particular also provide problems of air ingress. They are very prone to slackening when disturbed even slightly, and although leakage outwards may not be evident, in areas of the system where internal pressure is negative with respect to atmospheric, air may leak in. Such air will form very small bubbles, which may not be obvious, but will drain bubble traps, cause airlocked pumps, lines, columns and optical detectors. In addition, it may lead to microbiological contamination. In complex systems with many valves and joints in PTFE it is important that these are assembled in a rigid system and boxed with a minimum number of external bulkhead ports to avoid movement. As a further precaution, in-line filters, of the pre-sterilized capsule type fitted with a 0.2-μm porosity hydrophobic membrane set into the capsule should be employed. These hydrophobic membranes will act as vents to release entrapped air without fear of microbial contamination or leakage of process liquors.

The preferred material for full-scale industrial equipment with pipe diameters of 1 cm and upwards is undoubtedly stainless steel. Particular care must be taken with the quality of internal finishes and welds, as persistent microbiological problems may be caused by uncleanable working parts or crevices, particularly in pumps and valves. Sanitary pumps should be selected with all welds ground off and polished and all inner surfaces electropolished. In addition, some pumps are designed with internal clearances to facilitate cleaning and steam-sterilization.

6. SOLIDS SEPARATION

The main operations used for the separation of microbial and mammalian cells from their supernatants are centrifugation, conventional filtration (also called 'dead end filtration', given the operational procedure usually used) and membrane filtration (microfiltration and ultrafiltration) also called 'cross-flow filtration'. Dead end filtration is typically not successful for separation of biological suspensions. Comparison of centrifugation and filtration methods must take into account capital, operating and maintenance costs, as well as recovery yield and processing time. The overall yield of active product is often a function of the processing time and other factors such as the processing temperature.

One of the main factors influencing economics of separation processes is the size of the particles to be separated. As particle size increases, separation costs via centrifugation are substantially decreased, while filtration costs are less dependent on size

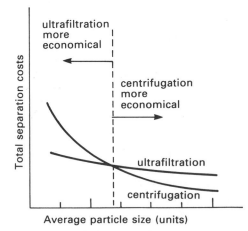

Figure 1. Centrifugation costs depend more on particle size than do membrane filtration costs (adapted from ref. 10).

(*Figure 1*). The cross-over point is in the $1-2$ μm range. Microfiltration will produce a sterile stream which is important for containment (and which centrifugation cannot produce), but cell fragments and debris will not necessarily be retained.

6.1 **Centrifugation**

A number of different centrifuges are available, all following the same physical principles for cell separation. The rate of settling for spherical particles, U, is given by (9):

$$U = \frac{r\omega^2 d_p \ (\varrho_p - \varrho_l)}{18\mu}$$

where r is radius of rotation, ω is angular velocity, d_p is particle diameter, ϱ_p is density of particles, ϱ_l is density of the liquid, and μ is viscosity. A correction factor is usually used for hindered settling for high particle densities. The expression shown is important in understanding the variables that affect centrifuge design: principally density difference between the cells and the liquid, diameter of the cells, and viscosity of the liquid.

Industrial centrifuges are generally of continuous flow through design and the faster the cells settle in the centrifuge, the shorter the necessary residence time in the rotor and the greater the centrifuge's processing capacity which has a profound effect on the economics of the process. One of the major factors controlling centrifuge capacity for biological separations is the size of the particles involved. Recovery of yeast has been carried out for many years by centrifugation; however yeast cells are ten times larger than bacteria thus for bacteria the centrifuge capacity is reduced by 100-fold. Mammalian cells present a further complication in that they are very fragile and are likely to disintegrate when passed through such a centrifugal system giving rise to finely divided debris.

The major centrifuge manufacturers have developed new machines specially for bacterial cell applications mainly by increasing the rotational speed (typically resulting in 14 000 $-$ 15 000 g). The most common centrifuge for the large-scale separation of bacteria, as well as cell debris, is the disc stack centrifuge. A properly-operated large-scale disk centrifuge should separate 99% of the solids from the liquid stream and produce an 80 $-$ 90% wet solids concentrate. Particles down to 0.5 μm can be separated with low flow rates of 300 $-$ 500 litres h^{-1}, which is sufficient to remove cell debris from suspension. To separate whole cells from suspension with this machine flow rates of 3000 $-$ 5000 litres h^{-1} can be used.

Some of the important limitations for centrifuge operation revolve around the generation of heat, aerosols and noise, as well as the lack of containment and the difficulty of sterile operation. This is particularly true for the very high-speed machines. The higher speeds lead to shorter residence times, but also lead to greater heat evolution and possibly an unacceptable temperature rise. To overcome this it may be necessary to arrange cooling of both the feedstock and the machine, an action which has cost implications. Aerosols are generated either by air entrainment through the centrifuge seals or by the use of compressed air for actuating the solids discharge mechanism. These aerosols must be contained at all times and particularly so in the presence of

9

toxic compounds and recombinant cells. Containment implies protection of the environment from biological materials in aerosol form generated within the machine. This has particular implications for the safety of the operators working the process. It also implies that the product stream within the centrifuge is protected from the outside environment and the microbiological contamination that this implies. Contained centrifuge systems are commercially available, fully hermetically sealed and provided with sterile filters (10).

6.2 Filtration

The separation of solids from liquids by filtration ranks amongst the most familiar of procedures in the chemical and biological sciences. The consideration of filtration procedures progresses more or less logically from depth filters to microporous filters to ultrafilters.

6.2.1 Depth filtration

Although centrifugation is the most widely used cell separation method, it is often costly. Other techniques that have been used for cell separation fall into the area of filtration, and amongst these the rotary vacuum filter has been the most popular one. Traditional depth filtration methods used in bioprocesses also included filter presses, but these are not used in modern biotechnology process design. Various types of rotary vacuum filters have been used. They are mainly used to remove bacterial and filamentous microorganisms from fermentation broths in large-scale enzyme and antibiotic plants. In conventional filtration, feed slurry impacts directly on to the filtration media, with all the mother liquor removed through the deposited solid cake and the filtration media. In most applications the filtration medium is a thick layer (5−10 cm) of precoat (for example siliceous earth) deposited on a rotating drum under vacuum from the inside. A thin layer of this plus the thick layer of cell deposit is removed by a knife in every revolution. Typical feed rates of fermenter broths are $100-200$ litres $m^{-2} h^{-1}$. Filter sizes are up to 100 m^2 filter area for the largest production units.

The utility of depth filters in large-scale processing of protein solutions lies in their ability to collect relatively large amounts of suspended solids like cell debris, hazes of denatured protein and ill-defined precipitates prior to either further filtration or chromatographic processing. A hazard is that the material of construction of the filter sheets may be a very good adsorbent for the target protein itself, and because of both the relatively large area of the filter and the large internal area represented by the surfaces of the fibres, losses during filtration may be unacceptably high. These problems should be addressed, firstly via the manufacturer's literature and representative, and secondly at the laboratory bench.

6.2.2 Membrane filtration

Membrane filtration systems are increasingly being used to separate whole cells and cell debris from cell suspensions, fermentation broths and cell homogenates. Removal of bacterial cells and cell debris from fermentation broths at pilot-plant scale has been the major application of microfiltration processes in biotechnology. They are particularly suited for containment and sterile separations, better than disc stack centrifuges which may also be more expensive. Membrane filtration also appears to offer lower operating costs than rotary vacuum filtration, mainly since the need for expensive filter aids is eliminated (11). A large variety of materials has been used for the manufacture of filtration membranes. These include, for microfiltration, porous thermoplastics (PTFE, nylon), ceramic (inorganic oxides such as aluminium oxide) and sintered metal (10,12), and for ultrafiltration, polysulphones and polyacrylamides. Configurations of membrane units can be tubular, spiral sheet (spiral wound), flat sheet and hollow fibre (1); all have been used in large-scale operations. Typical flux rates for cross-flow operation are $15-50$ litres $m^{-2} h^{-1}$ (11). Operation of the system in a cross-flow mode (12) seems to have a strong effect in increasing filtration rates by lowering concentration polarization on the membrane surface. The main disadvantage of membrane filters seems, however, to be the inability to handle a high concentration of cells, which dramatically lowers the filtration rate. It has thus been proposed to use a 'cascade' of three-cross flow units in series, the first producing a 10% solids suspension, the second a 20% solids suspension and the third a 30% one (13). Higher concentrations are virtually impossible to handle practically.

Microfiltration will retain very fine particulate material but not proteins in solution, whereas ultrafiltration will be able to retain proteins in solution. Microporous membranes used in microfiltration can retain particles as small as 0.01 μm. They are usually absolute filters, which means they will retain virtually all particles above a particular size. Typical filter sizes used in biotechnology to retain bacteria are 0.22 or 0.45 μm. These filters will produce sterile streams.

Scaling-up filtration systems must take into account an increase in area to permit the required flow rate and some consideration of the expected microbiological challenge. On the industrial scale, any interruption in the process flow in order to change a filter has serious cost implications, and a substantial margin of error should be allowed. An appropriate strategy is to remove suspended debris by means of a series of filters of progressively smaller pore sizes (for example 1.0, 0.45, 0.2 μm cut-off). Although there may be a statistical chance that microorganisms pass the first and even the second filter, the chances of passing all three are virtually zero, particularly if they are appropriately sized.

6.2.3 *Filtration problems*

(i) *Protein flocculation.* Most of the above applies to cohesive irregular particles such as cell fragments, or particles which are essentially small and solid like bacterial cells or spores. It is important to remember that denatured proteins readily flocculate to particles which are a great deal less well defined. Vigorous pumping through a filter will often break up these flocs, which will pass through very fine filters only to reform downstream. Under the much more gentle conditions encountered while loading on to a chromatographic column the flocs then collect on the much coarser in-line filter, causing unexpected blockages and potentially large spills. It is strongly recommended that protein solutions are filtered very gently and loaded on to columns by-passing in-line filters.

(ii) *Protein binding.* Proteins are surface-active and are prone to adsorb at many different kinds of interfaces. Filters, whether of the depth or microporous membrane type, have very large internal surface areas accessible to the protein molecules and, not surprisingly, it is necessary to take care to ensure that there is no appreciable loss of active product due to non-specific binding to this surface. Such binding is strongly dependent on the polymer concerned, but is also somewhat dependent on the individual protein species. As in any such situation, it is suggested that the filter manufacturers should first be consulted, followed by small-scale model experiments with the target protein under appropriate conditions of, for example, ionic strength, pH and temperature.

(iii) *Denaturation*. Conformational changes, which may represent profound changes in the protein to the extent of destroying biological activity, can be induced by filtration. Although it is tempting to attribute these to mechanical action, the damage seems to be more closely related to the surface denaturation seen in foams, in that the protein appears to be temporarily adsorbed to a surface, partially unwinds or otherwise changes conformation, and then is desorbed but does not return to its original native state. Again, this must be minimized by choice of filter medium with minimal protein adsorbtion.

7. CELL DISRUPTION AND DEBRIS SEPARATION

Cell disruption is necessary for the isolation of intracellular proteins. Recent developments in recombinant DNA technology have resulted in the cloning of many heterologous proteins in bacteria and yeast. For some of these host cell lines, secretion mechanisms have been found; however, for a substantial number this has not been possible. Hence, efficient and selective disruption techniques are an important step in producing many rDNA proteins.

A wide range of disruption techniques is used in the laboratory, both chemical (alkali, enzyme or detergent treatments) and mechanical (sonication, use of a pressure cell, homogenizer or bead beater). The chemical methods present problems of protein damage or contamination, and of the mechanical methods, the techniques most suitable for the large scale are the homogenizer and the high-speed bead mill.

7.1 Homogenization

In homogenizers, such as the Manton-Gaulin type, a cell suspension of up to $15-20\%$ dry weight is forced under pressures of up to 8000 psi through a very narrow annulus or valve. The almost instantaneous pressure drop produces effective disruption of the cells. Because of the large amounts of mechanical energy dissipated, temperature rises of up to $15\,^\circ$C can occur during a single pass. In order to overcome this, the feed-stream is usually precooled and cooled again after passing through the machine. Multiple passes, with cooling between, are often necessary to obtain the desired level of protein release.

The cell debris produced by this mechanical breakage has a high proportion of small fragments (14), and removal by continuous centrifugation is difficult because throughput is inversely related to the square of the particle diameter. Filtration is also made difficult, not only because of the high concentrations involved but also because of the gelatinous nature of the homogenate (7). In addition, some proteins will be denatured by the heat generated unless the device is very efficiently cooled. Although shear itself does not seem to be very detrimental to proteins, there is a tendency to generate foams by degassing or by the inclusion of air and shear at gas−liquid interfaces which can be very harmful (5). The largest available homogenizers can handle up to 6000 litres h^{-1}. Homogenization is at present the most widely used large-scale disruption technique; however, it is not suitable for some highly filamentous microorganisms.

7.2 **Bead milling**

In a bead mill a cell suspension is agitated at a high rate in the presence of small glass beads (0.3−0.4 mm diameter). This technique is particularly useful for the disruption of highly filamentous microorganisms.

An important difference from homogenization is that the concentration of cells will affect the efficiency of disruption, although the effect is smaller at higher agitation speeds. The optimum concentration of cells will vary depending on the microorganism, but is usually 30−60% cells wet weight (10). As in the case of the homogenizer, a substantial amount of heat is generated by the operation, and this is generally removed by means of a cooling jacket as there is normally no recycle stream. The largest available bead mills can handle throughputs of up to 2000 litres h^{-1}.

7.3 **Chemical and enzymic lysis**

Although protein release from microorganisms on the large scale is usually accomplished by mechanical disruption, such methods have several drawbacks. Since cells have to be broken completely to obtain a high yield, all intracellular materials are released. Hence the product must be separated from a complex mixture of contaminants (proteins, nucleic acids and cell wall fragments). In addition, nucleic acids will considerably increase the viscosity.

Chemical and enzymic cell permeabilization and lysis have not been used widely for large scale intracellular product release and disruption of microbial cells. However, both techniques are being investigated for this purpose (7,15−17) particularly as they offer the very attractive potential of differential protein release from different cell compartments (Huang,R. *et al.*, unpubl.) as well as selective membrane permeabilization (7) and the release of recombinant proteins (18,19).

Lysozyme is a commercial lytic enzyme that has been used for a number of years in industry (20) for different large-scale purposes, including enzyme extraction. However, this enzyme is active only on bacterial cells. Other bacteriolytic and yeast lytic enzymes (such as Zymolyase and enzymes from *Oerskovia* and *Cytophaga*) are available only as laboratory reagents, so it is not possible to make a meaningful cost comparison with other release methods on an industrial scale. Even so, design calculations based on enzyme production data in a 1000-litre fermenter and recent work on regulation of enzyme synthesis (21,22) demonstrate the economic viability of this approach (16).

Several chemical methods have been employed to extract intracellular components from microorganisms by permeabilizing outer wall barriers. An excellent review of this topic, which includes bacteria, yeast and mammalian cells, is that by Naglak *et al.* (7). Most work has been carried out with Gram-negative bacteria (mainly *E.coli*) using chelating agents (such as EDTA), solvents (such as 5% toluene), anionic and non-ionic detergents (such as sodium dodecyl sulphate and Triton X-100), and chaotropic agents (guanidine and urea). One of the very few large-scale applications of this technique

13

involved Gram-positive cells, where 0.5% Triton was used to release about 50% of the cholesterol oxidase from *Nocardia* cells (23).

Mammalian cells can be permeabilized using the steroid glycoside of digitonin (digitin) and the related saponins.

7.4 **Debris separation**

Centrifugation has been used for the separation of cell debris after disruption, but it is not a very good choice as it cannot satisfactorily handle particles with a diameter less than 0.5 μm (Section 6.1). This necessitates extremely low flow rates, even with machines producing very high centrifugal forces. Flocculation prior to centrifugation may improve the efficiency and economics; alternatively, cross-flow microporous membranes may be used. Even so, fouling of the membranes may limit the concentration of debris which can be achieved.

Two-phase aqueous partitioning or two-phase aqueous liquid-liquid extraction (24) is an operation which is gaining interest for the concentration of proteins and primary protein separation. In addition, this technique is extremely efficient in separating broken cells and cell debris which will partition to the lower phase from the proteins of interest which are made to partition to the upper phase.

8. PRIMARY SEPARATION OPERATIONS

Many modern biotechnological processes used for the manufacture of therapeutic proteins (for example monoclonal antibodies and other mammalian cell products) have a very low product concentration, perhaps around $10-100$ mg litre^{-1}. Therefore at an early stage in the purification process the protein must be concentrated $10-50$ times in order to lower the process volume and make the subsequent downstream processing more manageable. This is an important stage in an industrial process and, even if overlooked in the laboratory, it is essential on the large scale. A variety of techniques are in use to this end, although it is probably true to say that the preferred method is ultrafiltration.

8.1 **Ultrafiltration and diafiltration**

If the pore sizes are small enough, membrane filters can be used to concentrate proteins and peptides from dilute solutions. Ultrafiltration membranes have a much larger spread of pore size distribution than microfiltration membranes. They are rated in terms of their molecular weight cut-off, and on the large scale they are mainly used for concentrating biological solutions (reduction of water and salts content), or to separate materials with widely differing molecular weights (for example salts from proteins) since their molecular weight cut-off distribution is not particularly narrow and they produce quite a wide distribution of material (25,26). Typical molecular weight cut-offs used for concentration of proteins are 10 000, 30 000 or 50 000. Various manufacturers supply membranes described as having, for example, 10 000 or 20 000 dalton cut-off. These are figures based on statistical properties, determined using specific proteins and rounded off to the nearest $5000-10\ 000$. It is hence a mistake to treat these values as anything other than very rough guidelines.

Although it is theoretically possible to ultrafilter protein solutions directly, in practice the build-up of proteins on the surface of the membrane rapidly reduces the rate of flow dramatically. This 'concentration polarization' is controlled on the laboratory scale by agitation of the solution, but on the large scale a cross-flow system is more practicable. Membrane configurations are the same as used in microfiltration (see Section 6.2, and reference 1). The final concentration of protein may be very high; 25% solutions are possible on the large scale, but at this concentration there is virtually no flow through the membrane and the protein solution may become a 'solid gel'; this should be avoided. At any stage, the concentration effect may be stopped by continual addition of a more desirable buffer to the holding tank. As filtration continues, the composition of the protein solution will progressively approach that of the feed buffer. This technique of diafiltration can be useful in conditioning protein solutions in preparation for subsequent steps such as changing buffer or pH for ion-exchange chromatography.

8.2 **Precipitation**

Precipitation by neutral salts such as ammonium sulphate, organic solvents or other agents for recovering and purifying proteins is one of the oldest protein concentration methods known and is still commonly used for certain proteins (such as enzymes). However, precipitation methods are not favoured in most modern industrial biotechnology applications because the purification obtained with this technique is frequently minimal, and such methods introduce an extraneous agent which must later be eliminated and/or recovered. In addition, in very dilute protein solution, precipitation may proceed very inefficiently and result in poor yields (6). The use of organic solvents in particular on a large scale is generally undesirable, since considerable expense will be necessary for flame-proofing and spark-proofing equipment and plant areas.

8.3 **Aqueous two-phase extraction**

This technique, which is in many ways similar to the extraction of antibiotics using organic solvents, can be used for the separation of proteins mainly at laboratory and pilot-plant level; however, a few industrial processes do exist. Phase components will include water-soluble polymers and salts. Unlike non-polar solvents, they show low interfacial surface tensions and dielectric constants and hence they do not denature proteins (24). The best-studied systems for extraction of a number of proteins are the PEG-Dextran and the PEG-phosphate ones (9,27−30). They have been successfully used on a large scale, but applications are still limited. One technique that presents great potential is affinity purification or affinity partitioning, which is based on a biospecific ligand either present or bound to the PEG phase to increase selectivity (31,32). In general, large-scale application of two-aqueous-phase liquid extraction remains to be realized.

8.4 **Nucleic acids removal**

When intracellular proteins are released by cell breakage, nucleic acids are also released. This adds a contaminant which is not only difficult to separate but which also increases the viscosity of the suspension quite dramatically. Techniques used to remove nucleic acids include the use of nuclease enzymes and precipitation. A cellulose-based adsorber called cell debris remover or CDR (Whatman) has also been proposed as an appropriate method. CDR will remove cell debris as well as nucleic acids, lipids and polyphenols.

Most chromatographic steps will progressively reduce the level of nucleic acids in the product, but such contamination may still remain a problem in the case of therapeutic proteins due to the very low levels, measured in picograms per dose, demanded by regulatory authorities. Precipitation of nucleic acids is usually carried out in a rather efficient manner by using a long-chain cationic polymer with a molecular weight of about 24 000, called polyethyleneimine. This precipitation can be used to remove both nucleic acids and cell debris by centrifugation. A typical large-scale procedure is to use 1% of a 1% solution of polyethyleneimine, mixing thoroughly, when nucleic acids should precipitate almost instantaneously. Polyethyleneimine is usually not recovered.

9. PROTEIN PURIFICATION OPERATIONS

At this point one is faced with a 'broth' of proteins and some other components (perhaps some lipid material and/or wall or other polysaccharides), salts and water in a concentrated solution. After considerably reducing the solution volume in the previous concentration step, the protein content suitable for chromatographic purification is around $50-70$ g litre^{-1} total protein content (33). For the recovery, resolution and purification of a single protein, ideally one would like one step to extract virtually 100% of the protein from this mixture with no contaminants. However, typically two or, in some cases, three or four stages will be needed to achieve the final purity required for the particular application.

At this stage there will be a number of alternative combinations of different chromatographic steps (*Table 2*). The choice is often made empirically and confirmed by laboratory scale experiments. Computer methods used in classical chemical engineering process design, as well as expert systems are gaining favour for optimizing protein purification sequences (8), as will be discussed in more detail in Section 10 of this chapter. The principles of each type of chromatographic method are described in the companion volume, *Protein Purification Methods: A Practical Approach* (1) and hence are not discussed here.

Since most of the excess water has been extracted at this point, one would try to use a purification step of extremely high resolution in order to minimize the number of stages used, and hence maximize yield. However, this may not be possible in most cases at this stage, as some of the contaminants still present may produce fouling of the affinity or high-resolution ion-exchange column and hence shorten its life. Hence a first step in protein isolation for recovery of soluble proteins from other contaminants will probably be necessary. This would constitute a 'clean-up' step of pretreatment or primary isolation (see Section 3). For this, a relatively inexpensive treatment which should clarify the stream from suspended materials and non-protein contaminants in addition to salts should be used. This step will not give a very high purity but *must give a very high yield* in terms of our protein product recovered. Typical operations for this step would include inexpensive or disposable adsorption like a Whatman 52 ion-exchange cartridge, aqueous two-phase partitioning, or precipitation of the proteins using salt. After this step a high-resolution protein purification step will most probably be used. This stage should give a product of up to 99% (usually $95-98\%$) purity. Typical operations would be one or two high resolution ion-exchange chromatography steps or affinity chromatography. Although a high resolution is the main concern at this stage, an adsorbent that will *also* give a high yield should be chosen and/or designed.

Table 2. Chromatographic operations for large-scale purification of proteins.

Physico-chemical property	Operation	Characteristic	Use
Van der Waals forces H-bonds Polarities Dipole moments	Adsorption	Good to high resolution, good capacity, good to high speed	Sorption from crude feedstocks, fractionation
Charge (titration curve)	Ion-exchange	High resolution, high speed, high capacity	Initial sorption, fractionation
Surface hydrophobicity	Hydrophobic interaction	Good resolution, speed and capacity can be high	Partial fractionation (when sample at high ionic strength)
Biological affinity	Affinity chromatography	Excellent resolution, high speed and high capacity	Fractionation, adsorption from feedstocks
Isoelectric point	Chromatofocusing	Very high resolution, high speed and very high capacity (limited by size)	Fractionation
Molecular size	Gel filtration	Moderate resolution, low capacity, excellent for desalting	Desalting, end polishing, solvent removal
Hydrophilic and hydrophobic interactions	Reversed-phase liquid chromatography	Excellent resolution, intermediate capacity	Fractionation

Chromatofocusing has also been proposed as a very high-resolution operation to be used at this point on a relatively large scale.

After the high resolution step, a polishing step may be necessary to obtain ultra-high purity. This will depend mainly on the final use of the protein. If another physicochemical property cannot be exploited, gel filtration will be used which can separate dimers of the product (due to aggregation phenomena) or its hydrolysis products (due to action of proteases) solely on the basis of their different molecular weights.

For the purpose of scaling up, it is necessary to divide chromatographic processes into three classes: batch adsorption/desorption; isocratic chromatography and gradient methods. Chromatography is an extraordinary tool, and for the purification of a protein

Table 3. Effects of process parameters on resolution and throughput (35).

Parameter	Resolution varies with:	Throughput varies with:
Column length (L)[a]	L	$1/L$
Column radius (r)[b]	Some effect	r^2
Temperature (T)	Positive effect	T
Viscosity (η)	Negative effect	$1/\eta$
Sample volume (V)	$1/[V - V_{optimum}]$	V
Flow rate (J)	$1/[J - J_{optimum}]$	J

Bench-level purifications tend to be optimized only for resolution, while pilot and production scale purifications must also maximize throughput.

[a]For gel-filtration and isocratic elutions, column length is a critical factor. For gradient elution in adsorption chromatography, however, resolution is relatively independent of column length.
[b]Wall effects on resolution are pronounced in short-radius columns, and decrease as column length increases.

from crude preparations there are likely to be several chromatographic stages in any one process. Therefore careful consideration must be given to productivity throughout.

In virtually all chromatographic procedures, particularly in those based on adsorption-type interactions, scale-up is achieved by increasing the radius of the column while maintaining the column height. The effect of process parameters on resolution and throughput in chromatographic procedures is shown in *Table 3*. The resolution of proteins is mainly determined by the elution strategy (desorption) and not by the length of the column. Relatively short columns are favoured, in the range of 15−30 cm in height. Column capacity for large-scale use can easily be 100−150 litres. The largest available column for protein purification appears to be a 1700−2500 litre one, 2 m in diameter, with an adjustable bed height of between 55 and 80 cm (Amicon). A radically new column design constitutes radial flow chromatography (*Figure 2*) which has been developed by Sepragen as a more rational approach to scale-up chromatography (34). As shown in *Figure 2,* the mobile phase flows radially, and hence scaling up is achieved by increasing column length. The size of chromatography columns should be several-fold, for example 5−20 times (26), that of the adsorption capacity of the material as determined in batch experiments.

9.1 Batch adsorption/desorption

Batch adsorption/desorption will include any packed-bed process in which product is adsorbed on to the column and, perhaps after some intermediate conditioning or washing, is eluted in one step. Examples might be dye affinity chromatography, immobilized antibody columns or stepwise elution from ion-exchange columns.

The common primary consideration in all of these processes is that, in order to increase the scale, it is only necessary to increase the quantity of column packing. To scale 200-fold a 100-ml column will need 200×100 ml = 20 litres of packing. The problem then becomes a matter of geometry based on convenience. In general, column height is maintained while the diameter is increased. Flow through shorter, wider columns

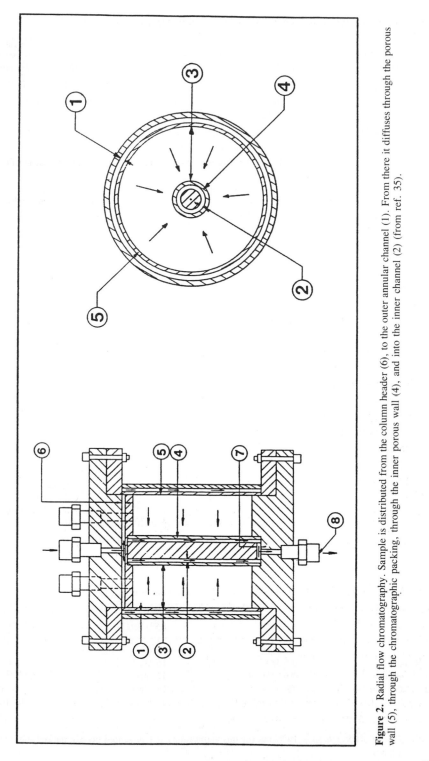

Figure 2. Radial flow chromatography. Sample is distributed from the column header (6), to the outer annular channel (1). From there it diffuses through the porous wall (5), through the chromatographic packing, through the inner porous wall (4), and into the inner channel (2) (from ref. 35).

is faster than through taller, narrower columns of the same volume, and the next criterion should be to ensure completion of this step within the time constraints imposed by other process stages. Over modest ranges, flow rate is inversely proportional to bed height, and proportional to the square of the radius. The extent to which these rules can be directly applied depends on the column packings. The highly cross-linked gels can be scaled over very wide ranges and the relationship between flow and geometry remains linear. Gels with a lower level of cross-linkage are more easily deformed. Increasing the column diameter beyond about 25 cm will remove the effect of wall support and can lead to gel compression and performance in terms of flow also becomes much less linear. For batch adsorption/desorption on Sepharoses (Pharmacia), for example, bed depths should be restricted to $15-20$ cm and flow rates scaled based on (radius)$^{1.5}$ rather than (radius)^2, so that an increase in diameter of 10-fold should be projected as an increase in flow rate of say 30-fold rather than 100-fold. These values look conservative, but it should be borne in mind that they must be practically attainable—the difference between laboratory science and engineering referred to earlier. An example of scale-up of an ion-exchange method is given in *Table 4*.

9.2 Isocratic chromatography

Isocratic chromatography, in the sense used here, refers to processes where product is loaded onto the column and is progressively eluted without change in buffer composition. This would include, for example, isocratic elution from an ion-exchange column and gel-filtration chromatography. In either case, the common primary consideration is that separation is directly dependent on column length. During scale-up, then, the column length must be preserved or increased in order to maintain performance, as resolution is proportional to the square root of bed height. In this case, scale-up should be envisaged as replicating laboratory columns in parallel, that is, the column should be designed to increase cross-section surface area in proportion to the required load, but to maintain height. As with a batch process, there may be a loss of efficiency on scale-up which suggests that a somewhat longer column is justified. This has penalties in slowing down an already tedious column step. Depending on conditions, this might be acceptable, or might be overcome by collecting a narrower cut from the product peak. Under some circumstances, it may be possible to increase throughput, and effectively reduce processing time by loading a second aliquot on to the column before the

Table 4. Scale-up of ion-exchange. Purification of albumin from plasma (33).

| | | Column | |
	K 50/30	*KS 370/15*	*GF 08-015*
Bed height	15.2 cm	15 cm	15 cm
Bed volume	0.3 litre	16 litre[a]	75 litre
Sample volume	0.64 litre	42 litre	150 litre
Linear flow rate	25 cm h^{-1}	22 cm h^{-1}	24 cm h^{-1}
Flow rate	0.49 litre h^{-1}	23.6 litre h^{-1}	120 litre h^{-1}
	1.6 V_t h^{-1}	1.5 V_t h^{-1}	1.6 V_t h^{-1}

[a]The volume of a KS 370/15 is 16 l, but for this application 20 l of settled gel are packed into the bed.

Table 5. Scale-up of gel filtration. Fractionation of albumin (33).

	Column	
	K 50/100	*KS 370/15 × 4*
Bed height	60 cm	60 cm
Bed volume	1.18 litre	64 litre
Sample volume	0.09 litre	5 litre
(0.078 x V_t)		
Linear flow rate	16.2 cm h^{-1}	11.25 cm h^{-1}
Flow rate	0.32 litre h^{-1}	12 litre h^{-1}
	0.27 V_t h^{-1}	0.18 V_t h^{-1}

first is completely eluted. Again, these are matters of judgement and specific cases. An example of scale-up of a gel-filtration method is given in *Table 5*.

9.3 Gradient methods

Gradient methods are not commonly scaled up by choice and should be avoided if possible. It is not easy to make reproducible gradients on the large scale, and a generally more satisfactory approach is to convert to a batch step elution process.

   If it is essential to use gradient elution on the larger scale, there is often a conceptual problem in understanding the parameters for scaling. Consider the example of a laboratory column of 1 cm^2 × 3 cm height eluted using a gradient of 20 mM to 100 mM in 15 bed volumes. Any particular gel particle has seen a gradient of 15 × 3/1 = 45 cm with a total height of 100 − 20 = 80 mM and a slope of 80/45 = 1.78 mM cm^{-1}. In order to scale this for a one-litre column (area 78.5 cm^2, height 12.7 cm) the gradient needs to reach the same height over the same fluid length (that is 80 mM over a length of 45 cm or a volume of 78.5 × 45 = 3.53 litres). The height of the column has little effect on the separation. The capacity of the column should be scaled on protein load per cm^2 of surface area in order to approach the original column conditions. Flow rates are maintained close to laboratory rates, based on litres per minute per area of bed surface.

9.4 HPLC and other methods

High performance liquid chromatography (HPLC) uses small rigid uniform beads (10−40 μm diameter) to produce separations with considerably higher resolution than conventional matrices. It is therefore often used for difficult separations requiring high resolution for success, such as purification of synthetic polypeptides to homogeneity (11). However, the matrices and equipment required are very expensive and the technique must be selected cautiously on economic grounds.

   A new methodology being presently developed for large-scale use is FPLC (fast protein liquid chromatography). FPLC systems will operate under moderate pressure, giving higher speed and resolution than traditional chromatography ones. Several purpose-made column packings and prepacked columns are available. FPLC has had very wide acceptance in the laboratory, hence it shows promise for the pilot- and full-plant scale depending on the developments and availability of suitable equipment. An advantage over high-pressure systems such as HPLC is that the equipment is constructed from plastics which, unlike stainless steel equipment, resist all buffers likely to be employed in the separation of proteins.

9.5 Chromatography matrices

Any column packing material consists essentially of active sites mounted on a matrix. Active sites may be hydrophobic, ionic or simply porous depending on the particular chromatographic technique. The purpose of the matrix is simply to support the active sites and to have no other interaction with the protein or other solution components.

With this in mind, we can summarize the key desirable properties of the matrix as follows.

(i) *Rigidity*. This minimizes deformation of the gel itself which tends to close down the channels between particles. More rigid gels will have a more extended linear relationship between flow rate and pressure. Typically the more rigid matrices should be chosen for large scale applications. In addition, matrices containing ionic groups must resist the changing internal osmotic pressure with variation of charge (due to pH changes) and ionic strength.

(ii) *Low non-specific interaction*. This refers to the problem of adsorption of protein to the matrix by mechanisms other than that intended. This leads to difficulties in recovering product, tailing of peaks for instance, difficulties in regenerating the column to a satisfactory state, and ultimately a limited column life.

(iii) *Chemical stability.* This is particularly important in terms of resistance to sanitizing agents such as sodium hydroxide, and stability to autoclaving for critical applications.

(iv) *Open pore structure.* In the case of derivatized varieties, i.e. other than gel filtration media, this allows macromolecules access to active sites and minimizes gel filtration effects.

A variety of media is available commercially, the main ones of which are the following.

(i) *Cellulose.* A medium available for many years, cellulose is difficult to pack uniformly due to its fibrous form and even more difficult to regenerate *in situ* without repacking the column. The pore size is small, and non-specific adsorption is a problem. However, it is possible to exploit this property by using it as a clean-up step for dirty extracts resulting from clarification (cell separation) of broken cell extracts. Cellulose ion-exchangers adsorb lipids, nucleic acids, coloured materials and suspended solids. They are best used as a once-off disposable packing before a high-resolution purification. Whatman DE52, DE32 and CDR (cell debris remover) can all be used in this way.

(ii) *Cross-linked dextrans.* Cross-linked dextrans (Sephadex, Pharmacia) represent a range of early tailor-made matrices, with much improved non-specific adsorption properties and resistance to dilute sodium hydroxide solutions and to autoclaving. The lower cross-linked varieties, which have pore structures open enough to admit proteins, are very soft and prone to deformation. This leads to problems in use, when the height of the packed bed may be compressed and reduced markedly during elution. In order to use the less cross-linked range, G75 to G200, long columns (for example 1 m) must be built up in short sections which can then be linked together to produce the total overall length required. This apparently unpromising compromise in fact works well in practice, and a range of 15 cm high column sections have been developed by Pharmacia for this purpose (KS370 'Stack' columns).

Varieties of cross-linked dextrans with ionic groups introduced showed marked swelling and shrinkage effects with change of ionic strength and pH. As in the case of cellulose, these materials have found a specific niche in protein processing in the form of desalting by gel-filtration using the highly cross-linked forms (G-25). The process is rapid, robust, amenable to scale-up and easy to automate. In addition, the columns are easy to regenerate and not difficult to sanitize. In some cases, desalting by gel filtration is preferred to ultrafiltration.

(iii) *Agaroses.* Agaroses, such as Ultrogel A (IBF), Biogel A (Bio-Rad) and Sepharose (Pharmacia) and cross-linked agaroses [such as the Sepharose CL and Sepharose FF ranges (Pharmacia)], probably represent the most popular matrices for a wide variety of chromatographic techniques. Non-specific adsorption is low, almost as low as the cross-linked dextrans, and cross-linked agaroses can be cleaned and sanitized with dilute sodium hydroxide and can be autoclaved if necessary. The open pore structure gives good access to proteins in the derivatized forms, of which there are many, and the more rigid nature leads to very satisfactory flow properties in packed columns.

(iv) *Polyacrylamide.* A variety of column packing materials is available based on polyacrylamide, for example Biogel P (Bio-Rad) or hybrids of cross-linked polyacrylamide with agarose [Ultrogel AcA range (IBF)], or with dextran [Sephacryl, (Pharmacia)]. These tend to combine the chromatographic virtues of their components, but also the disadvantages in terms of non-specific adsorption and lack of resistance to dilute sodium hydroxide and autoclave conditions. They are used on the large scale, but are perhaps best reserved for critical and problematic separations where more complex santitation or regeneration routines can be justified.

(v) *Others.* For completeness, one should mention that other column packings are used for protein purification in very specific cases, including controlled-pore glass (Corning) and derivatized silica (Vydac). Both of these are used for specific purification projects on the large scale, but these applications are few.

10. PROCESS DESIGN, OPTIMIZATION AND ECONOMICS

In Section 3 of this chapter, the choosing of a potentially sound large-scale purification strategy with a minimum number of steps and giving a high yield was discussed. The optimization of a process design is carried out after one or more purification strategies have been chosen and performed, and thus process conditions are known as well as the characteristics of the final product. The interest at this point is to find the optimal operating conditions for specific separation operations so that it is possible to compare its performance with alternative operations. It would also be desirable to optimize performance and compare alternative chromatographic sequences (giving a similar final product) in terms of overall economics (36).

10.1 Design parameters

It is important to have information on the physicochemical characteristics of the final protein product and of the main contaminants (see *Tables 6* and *7*; 37). This will have an impact on the level of rigorousness that can be used in the analysis of the interaction

Table 6. Typical properties of a desired protein (37).

Primary amino acid sequence

Isoelectric point (electrical mobility)

Solubility characteristics:
  pH
  Buffer systems
  Organic solvents

Stability characteristics:
  pH
  Temperature
  Buffers
  Hydrophobicity and hydrophilicity characteristics

Sedimentation and diffusion coefficients

Location of protein:
  Cytoplasmic
  Periplasmic

Table 7. Typical properties of contaminating materials (37).

DNA and RNA:
  Contents
  Properties

Major protein contaminants:
  pI
  Solubility
  Amount
  Electrical mobility

Membrane components
  Contents
  Properties
  Solubility

---

of the protein and contaminants with the individual separation processes.

Secondly, quantitative and numerical information on the performance of individual operations is vital. For instance, in the case of mechanical cell disruption, the design information available has been described in the literature (14). It will include flow rate, type of agitator (or operating pressure), cell concentration and type, fraction of product release and size of the disrupter. For a chromatography operation the information necessary has been described mainly in Section 9. Descriptions are given in several sources on the theoretical and design analysis of adsorption type chromatography (1, 38 − 42); it concerns characteristics of columns (size, geometry) and properties of gels and other adsorbents (binding capacity, dissociation constants, flow rate, half-life, breakthrough curves).

Thirdly, mathematical models and mathematical correlations of the operations can be used (38−42). This will allow simulation of performance and also may be used to scale up individual operations. Computer simulations are a useful tool to optimize separations in chemical process engineering. Examples of useful downstream process simulations, and investigation of process conditions, are microbial cell breakage and selective product release using enzymes (17,18,19,43,44) and investigation of the affinity chromatography of proteins (45).

10.2 Process synthesis and optimization

The problems that have to be solved in process synthesis and optimization of downstream protein separations appear to be of two types: (i) choosing between alternative operations (e.g. homogenizer *v.* bead mill, chemical *v.* enzymatic cell permeabilization, or centrifugation *v.* cross-flow microfiltration), and (ii) the design of an optimal chromatographic sequence with maximum yield and minimum number of steps (one, two or three). These problems can be solved either by finding a rigorous solution using numerical methods, or by using an Expert Systems (ES) approach (8). The first solution by itself has limited use in biotechnology due to a lack of useful design equations and databases. The second approach appears more attractive, since it allows the use of empirical knowledge and extensive scattered information. Computer-based expert systems are an important tool in the field of artificial intelligence (AI). They allow the use of scattered information and databases, empirical knowledge, heuristics (rules of thumb) as well as established mathematical correlations and strict quantitative data. Existing AI systems use an interactive inferential logic to find an optimum solution to the problem of protein purification processes and sequences. Efforts have been made to develop AI systems for this purpose (46,47) or to adapt existing systems (called 'shells') (48) both for the manipulation of heuristics, databases and simple algebraic design equations.

10.3 Economics

The purpose of scale-up is two-fold; first, to produce more product, and second, to produce it more economically, by reduction of the cost per gram. In this section we have tried to look at some of the costs associated with running a hypothetical process at the laboratory (say 10 ml column), pilot (1-litre column) and production (25-litre column) scales. Based on these values, and assuming a three-column purification process with intermediate stages such as ultrafiltration, we can review costs at least on an approximate basis. All processes differ, and different columns in any overall process scheme will have different volumes, nonetheless the figures should be indicative.

We have divided costs into chemicals, consumables (plastic sample pots, sterile disposable pipettes, filter cartridges), capital (equipment, pumps, tanks), facility (the share of the cost of the building) and labour (in man-days). Note that labour will be charged at an 'overheaded' rate, not at bare salary; this will include some way of paying off the cost of running the facility, the cost of heating, air-conditioning and other highly diffused costs. *Table 8* summarizes estimates of costs on this basis, including cost per unit column volume, which should be proportional to unit cost if we assume that quantity of product is proportional to column volume.

Table 8. Costs of scale-up.

Scale Ratio	10 ml 1:1	1 litre 100:1	25 litre 2500:1
Chemicals cost[a]	1	20	500
Disposables cost[a]	25	150	2500
Unit cost[b]	2.6	1.7	1.2
Equipment cost[a]	10 000	15 000	250 000
Unit cost[b]	1000	150	100
Man-days, production	5	7	15
Man-days, other	–	4	10
Total	5	11	25
Relative cost	0.5	0.011	0.001
Facility share	?	?	1 M

[a]Costs are relative to chemicals cost at 10 ml scale of 1 unit.
[b]Cost per unit column volume.

We can recognize the following from the table.

(i)   The cost of chemicals does not rise in direct proportion to batch size. At laboratory scale, solution volumes are chosen for convenience and a large proportion is discarded. At the pilot and production scale, volumes more closely reflect the needs of the process.

(ii)   Filters and disposable plasticware costs are not stable but rise to perhaps surprisingly large figures. There is much more sampling at the production scale and solutions will be filtered after each process step to minimize 'bioburden'. The unit costs of these operations are roughly linear with scale.

(iii)   Equipment costs rise dramatically with increasing scale. At the laboratory scale these represent a fraction collector, UV monitor, recorder and little else. At the pilot scale, stainless containers and some modest automation pushes the costs up somewhat, but a fully equipped production laboratory with hard-piped layout and jacketed stainless steel tanks with 'clean-in-place' facilities represents a substantial investment. In relation to batch size, however, the costs fall by an order of magnitude, in line with the primary purpose of scale-up.

(iv)   The table also shows a £1M share of the overall facility cost applied to the purification area. Clearly, for production there must be offices, a warehouse, laboratories and a fermentation hall in addition to a purification area. Some proportion of this overall cost must be applied as purification costs. No attempt has been made to assess this for other scales (this may be left to accountants), but it is important to recognize that it exists.

(v)   Finally, the total manpower requirements in man-days per batch rise sharply with scale; large batches mean more material to move around, more sampling and control activities and more documentation. In relation to batch size, however, costs fall by about two orders of magnitude in line with the primary purpose of scale-up.

11. REFERENCES

1. Harris,E.L.V. and Angal,S. (eds) (1989) *Protein Purification Methods: A Practical Approach*. IRL Press, Oxford.
2. Rudd,D.F., Powers,G.J. and Siirola,J.J. (1973) *Process Synthesis*. Prentice Hall, Englewood Cliffs, NJ.
3. Wheelwright,S.M. (1987) *Bio/Technology*, **5**, 789.
4. Wang,H.Y. (1983) *Ann. N.Y. Acad. Sci.*, **413**, 313.
5. Fish,N.M. and Lilly,M.D. (1984) *Bio/Technology*, **2**, 623.
6. Cartwright,T. (1987) *Trends Biotechnol.*, **5**, 25.
7. Naglak,T.J., Hettwer,D.J. and Wang,H.Y. (1989) In *Separation Processes in Biotechnology*. Asenjo,J.A. (ed.), Marcel Dekker, New York, in press.
8. Prokopakis,G.J. and Asenjo,J.A. (1989) In *Separation Processes in Biotechnology*. Asenjo,J.A. (ed.), Marcel Dekker, New York, in press.
9. Kula,M.R. (1985) In *Biotechnology*. Rehm,H.J. and Reed,G. (eds), *Fundamentals of Biochemical Engineering*, Vol. 2, Verlag Chemie Mannheim, Chap. 28.
10. Bjurstrom,E. (1985) *Chem. Eng.*, **95** (4), 126.
11. Brocklebank,M.P. (1987) In *Food Biotechnology*, King,R.D. and Cheetham,P.S.J. (eds), Vol. 1, Elsevier, Amsterdam, p. 139.
12. Brown,D.E. and Kavanagh,P.R. (1987) *Process Biochem.*, **22**, 96.
13. Hedman,P. (1984) *Int. Biotechnol. Laborat.*, May/June, 1984.
14. Schutte,H., Kroner,K.H., Husted,H. and Kula,M.R. (1983) *Enzyme Microb. Technol.*, **5**, 143.
15. Hettwer,D.J. and Wang,H.Y. (1986) In *Separation, Recovery and Purification in Biotechnology*. Asenjo,J.A. and Hong,J. (eds), ACS Symp. Ser., Vol. 314, American Chemical Society, Washington DC, p. 2.
16. Andrews,B.A. and Asenjo,J.A. (1987) *Trends Biotechnol.*, **5**, 273.
17. Hunter,J.B. and Asenjo,J.A. (1986) In *Separation, Recovery and Purification in Biotechnology*. Asenjo,J.A. and Hong,J. (eds), ACS Symp. Ser., Vol. 314, American Chemical Society, Washington DC, p. 9.
18. Asenjo,J.A., Andrews,B.A. and Pitts,J.M. (1987) *4th Eur. Congr. Biotechnol.*, Amsterdam, June, 1987.
19. Asenjo,J.A., Andrews,B.A. and Pitts,J.M. (1988) *Ann. N.Y. Acad. Sci.*, **542**, 140.
20. Scott,D., Hammer,F.E. and Szalkucki,T.J. (1987) In *Food Biotechnology*. Knorr,D. (ed.), Marcel Dekker, New York, p. 413.
21. Andrews,B.A. and Asenjo,J.A. (1986) *Biotechnol. Bioeng.*, **28**, 1366.
22. Andrews,B.A. and Asenjo,J.A. (1987) *Biotechnol. Bioeng.*, **30**, 628.
23. Buckland,B.C., Richmond,W., Dunnill,P. and Lilly,M.D. (1974) In *Industrial Aspects of Biochemistry*. Spencer,B. (ed.), Vol. 30, Part I, Elsevier, Amsterdam, p. 65.
24. Andrews,B.A. and Asenjo,J.A. (1989) In *Protein Purification Methods: A Practical Approach*. Harris,E.L.V. and Angal,S. (eds), IRL Press, Oxford, p. 162.
25. Scopes,R.K. (1987) *Protein Purification*. 2nd. ed, Springer, New York.
26. Harris,E.L.V. (1989) In *Protein Purification Methods: A Practical Approach*. Harris,E.L.V. and Angal,S. (eds), IRL Press, Oxford, p. 125.
27. Dove,G.B. and Mitra,G. (1986) In *Separation, Recovery and Purification in Biotechnology*. Asenjo,J.A. and Hong,J. (eds), ACS Symp. Ser., Vol. 314, American Chemical Society, Washington DC, p. 93.
28. Husted,H. (1986) *Biotechnol. Lett.*, **8**, 791.
29. Albertson,P.A. (1986) *Partition of Cell Particles and Macromolecules*. 3rd. edn, John Wiley, New York.
30. Joelsson,M. and Johansson,G. (1987) *Enzyme Microb. Technol.*, **9**, 233.
31. Head,D.M., Andrews,B.A. and Asenjo,J.A. (1989) *Biotech. Techniques*, **3**, 27 (Paper presented at Int. Conf. Separations for Biotechnology, Reading, Sept. 1987).
32. Birkenmeier,G., Kopperschlager,G., Albertson,P.A., Johansson,G., Tjerneld, F., Ackerlund,H.E., Berner,S. and Wickstroem,H. (1987) *J. Biotechnol.*, **5**, 115.
33. Pharmacia (1983) *Scale Up to Process Chromatography. Guide to Design*. Pharmacia AB, Uppsala, Sweden.
34. McCormick,D. (1988) *Bio/Technology*, **6**, 158.
35. Sofer,G. and Mason,C. (1987) *Bio/Technology*, **5**, 239.
36. Duffy,S.A., Moellering,B.J. and Prior,C.R. (1988) Paper presented at the 196th ACS National Meeting, MBTD division, 25−30 Sept. 1988, Los Angeles.
37. Wang,D.I.C. (1987) In *Separations for Biotechnology*. Verrall,M.S. and Hudson,M.J. (eds), Ellis Horwood, Chichester, p. 17.
38. Yamamoto,S., Nomura,M. and Sano,Y. (1987) *AIChE Journal*, **33**, 1426.
39. Yamamoto,S., Nakanishi,K. and Matsuno,R. (1988) *Ion-Exchange Chromatography of Proteins*. Marcel Dekker, New York.

40. Treybal,R.E. (1980) *Mass Transfer Operations.* McGraw-Hill, New York, Chap. 11.
41. Ladisch,M.R. and Wankat,P.C. (1988) In *The Impact of Chemistry on Biotechnology.* Phillips,M. *et al.* (eds), ACS Symp. Ser., Vol. 362, American Chemical Society, Washington DC, p. 73.
42. Wang,L. (1989) In *Separation Processes in Biotechnology.* Asenjo,J.A. (ed.), Marcel Dekker, New York, in press.
43. Hunter,J.B. and Asenjo,J.A. (1988) *Biotechnol. Bioeng.,* **31**, 929.
44. Liu,L.C., Prokopakis,G.J. and Asenjo,J.A. (1988) *Biotechnol. Bioeng.,* **32**, 1113.
45. Chase,H.A. (1988) Paper presented at the Symposium on Antibodies for Purification, SCI, London, March 1988.
46. Siletti,C.A. and Stephanopoulos,G. (1986) Paper presented at the 192nd ACS National Meeting, Anaheim, CA, September 1986.
47. Wacks,S. (1987) *Design of Protein Separation Sequences and Downstream Processes in Biotechnology; Use of Artificial Intelligence.* M.Sc. Thesis, Columbia University, New York.
48. Asenjo,J.A., Herrera,L. and Byrne,B. (1989) *J. Biotechnol.,* in press.

CHAPTER 2

Purification of proteins for therapeutic use

A.F.BRISTOW

1. INTRODUCTION

1.1 Therapeutic uses of purified proteins

Purified and partially purified proteins have been used in the treatment and management of a wide range of disease states. This chapter will consider the range of proteins used in therapy, together with the range of therapeutic regimens that are used, the types of contamination that may occur in proteins for therapeutic use and the possible clinical consequences of the various types of contaminant. At a practical level, the section will consider analytical techniques used to assess the level of specific contaminants and the design and validation of purification processes for therapeutic proteins.

Therapeutically useful proteins may be divided into four groups: regulatory factors (including hormones, cytokines, lymphokines and other regulators of cellular growth and metabolism); blood products (including serum-derived blood factors, and enzymic fibrinogen activators); vaccines; and monoclonal antibodies. The four classes of protein exhibit a wide range of biochemical structures. Cytokines and other regulatory factors include tripeptides such as thyroliberin, small well-characterized proteins such as insulins, growth hormone and interferons, and larger, more complex molecules such as erythropoietin and gonadotrophins, which may be both glycosylated and multi-subunit. Most proteins in the blood products category are complex, varying from enzymes such as urokinase which may exist in a variable ratio of high- to low-molecular-weight forms, to high-molecular-weight assemblies such as the clotting factor VIII complex. Vaccines are generally high-molecular-weight structures and may exist, as in the case of hepatitis B vaccine, as a glycoprotein complexed with a lipid component. Monoclonal antibodies are generally IgGs, although IgMs may also be considered for therapeutic use. It can be readily appreciated that with the advent of recombinant DNA technology, the range and quantity of proteins available to the clinician has increased enormously, and the situation now exists that more or less any protein considered likely to be of possible therapeutic benefit in man can be made available in sufficient quantity for clinical evaluation.

In attempting to estimate the level of purity required for any of these proteins to be administered to man, several factors must be taken into account. Of prime importance is the source of the protein, since the presence or absence of specific impurities depends upon its route of manufacture. Traditionally the source would have been extracted animal tissue or human serum. Sources now include cultured microorganisms, mammalian cells infected with various viruses, a whole range of both prokaryotic and eukaryotic recombinant host—vector expression systems, and hybridomas of murine and possibly even human origin. The specific contaminants likely to be encountered will depend to a large extent on which of these systems is used for the production of the desired protein. Proteins may be administered to man in a wide variety of different therapeutic regimens. Insulin for example is given two or three times daily for a lifetime, compared with a vaccine which may be given once to three times only. Products such as interferons may be given for limited periods of intensive therapy whilst tissue plasminogen activator or monoclonal antibodies may be given as single-shot therapy in very high doses. Such factors need to be borne in mind when deciding upon the level of purity required. For instance, there need be less concern about the possibility of immunogenic contaminants being present in a preparation that is to be given once or twice only compared with a preparation that is to be given as daily replacement therapy for life. Finally, the nature and severity of the illness being treated is also of some relevance in deciding upon the acceptable levels of contamination. Medicines that are being used to treat or to vaccinate against non-life-threatening conditions might reasonably be expected to have to satisfy far stricter safety requirements than those used to treat life-threatening or otherwise incurable conditions. In the treatment, for example, of patients who are immune-deficient, as a consequence either of their illness or of therapy, one would be far more concerned about the possible contamination of the protein with non-human infectious agents (e.g. murine viruses derived from monoclonal antibody-producing hybridomas) than one would in the immune-competent patient where such agents would not normally be infective. It is clear therefore that to a certain extent each case should be considered on its own merits. Despite this, it is still possible to develop an over-all approach to the problem of purifying proteins for therapeutic use.

1.2 Methods included in this chapter

Detailed methodology for protein purification will not be given in this chapter, since all the relevant techniques have been covered in the companion volume, *Protein Purification Methods: A Practical Approach* (18). This chapter will instead provide details of analytical procedures used in assessing the levels of various classes of contaminant in proteins for therapeutic use. Most of the techniques described in detail will relate specifically to proteins produced in *E. coli*. All of the methods are applicable, with appropriate modifications, to proteins produced in other recombinant systems, or extracted from human or animal tissues.

2. CLINICAL SIGNIFICANCE OF CONTAMINANTS

2.1 Antigenicity

When a foreign protein is injected into man, it will normally generate an immune response. We may consider two types of immunity that may arise. The patient may become immune either to contaminants in the preparation which in the case of most recombinant expression systems will be non-human proteins and would be expected to be immunogenic, or the patient may become immune to the therapeutic protein itself, an outcome which is less likely since, for the most part, they will be human proteins and therefore non-immunogenic. There are however two mechanisms by which impure preparations may give rise to immunity to human proteins: degraded or aggregated forms of the product may be present which are more immunogenic than the native form;

and high levels of contaminating proteins from the host cell in a recombinant DNA expression system may act as adjuvant-like material, rendering the product protein more immunogenic than it would otherwise be. Immunity either to contaminants or to the major protein may have adverse clinical consequences. Immunity to foreign proteins may cause hypersensitivity upon repeated injection of similarly contaminated preparations, and it is obviously clinically undesirable to render patients autoimmune to their naturally occurring proteins.

2.2 Transformation

Concern also exists that contaminating DNA may be capable of causing transformation. Many cell lines used in production of proteins by recombinant DNA technology are known to contain active oncogenes, and these present a theoretical hazard, particularly in patients that are immune-deficient. Although little evidence is available on the ability of naked oncogenic DNA to induce or promote tumours, either in normal or in immunosuppressed animals, contaminating DNA derived from host−vector systems that use tumorigenic cell lines must be regarded as potentially clinically significant.

2.3 Transmissable disease

Proteins derived from a number of sources are potentially contaminated with agents either known to be pathogenic in man, or perhaps capable of being pathogenic in immune-deficient patients. Proteins extracted from human serum or tissue are potentially contaminated with the causative viruses for AIDS, hepatitis or Creutzfeldt-Jacob disease, whilst proteins derived from human cells in culture may be contaminated with a range of human viruses, including transforming viruses such as Epstein-Barr virus. A number of murine viruses may contaminate hybridomas used to produce monoclonal antibodies, some of which, such as Hanta virus, are known to cause serious, life-threatening disease in normal humans (1), whilst many others must be considered potentially capable of causing disease under some circumstances.

2.4 Pyrogenicity

Bacterial endotoxins are lipopolysaccharide components of the cell walls of mainly Gram-negative bacteria and are potential contaminants of any reagent used in the manufacture of proteins, particularly of bulk reagents such as water. They also represent a particular cause of concern in recombinant DNA expression systems using Gram-negative bacteria. Pyrogen contamination is clinically relevant, since administration of small quantities to man may result in a range of acute phase effects, including pyrexia, or fever (2).

3. ANALYTICAL TECHNIQUES

3.1 Detection limits

In order to determine the level to which proteins for therapeutic use need to be purified, and indeed, what sort of analytical techniques should be employed in order to test for the presence or absence of contaminants, we should consider what levels of contaminants are likely to cause the kind of clinical consequences that we have foreseen in the previous section. There has been considerable experience in the long-term administration of various preparations of insulins which are contaminated to different degrees with both insulin-related and non-insulin-related proteins, and it would seem that the limit for contaminants, below which antibodies are not formed, is of the order of 10 parts per million (3). Naturally this figure will vary according to the therapeutic dose and one may also expect different contaminants to exhibit different intrinsic immunogenicities. Nonetheless, the figure of 10 p.p.m. is a useful guide-line. As far as DNA is concerned, there is far less evidence as to what constitutes a hazardous dose in man. Figures of 10 or 100 picograms per therapeutic dose have been adopted, although again, the level of concern might be expected to vary with the source of the DNA. The limit of contamination with bacterial endotoxins is set at around 35 ng per therapeutic dose. With disease-causing viruses such limits are difficult to set, since in theory at least a single virus particle is capable of causing infection, and the only reasonable solution is to demonstrate that such viruses are absent.

It can be readily appreciated that measuring these extremely low levels of contaminants requires a rather more stringent approach than most other applications mentioned in this book. Generally speaking, the purification technique used is often used as the analytical tool to demonstrate purity, thus a protein purified by HPLC would be shown to be pure using the same technique. Detection limits, however, for such chromatographic procedures would be expected to be of the order of 0.1% or 1000 p.p.m., considerably above the limits for proteins or DNA previously discussed. Thus, whilst it is obvious that proteins for therapeutic use will have to satisfy conventional chromatographic and electrophoretic criteria of purity, one must also consider a completely separate battery of analytical techniques for measuring specific contaminants of clinical relevance. It is to these techniques that the practical aspects of this chapter will be addressed.

3.2 Immunological methods

Immunological methods depend on the production of specific antisera raised against possible contaminants which are then used for either quantitative or qualitative determination of those contaminants.

The variety of techniques available to the analyst is wide, and the reader is referred to general texts on practical immunochemistry, such as Johnstone and Thorpe (ref. 4). The specific techniques given below are intended to serve as examples, and many refinements are possible.

3.2.1 *Production of antisera*

In the case of a protein produced in a recombinant DNA system, such as *E. coli*, antisera recognizing many host cell proteins may be raised against a lysate of the entire host—cell (*Method Table 1*).

Note that, in the example given, the antigen preparation is produced by detergent solubilization and hence denaturation of the host cell proteins. Such procedures are appropriate where the product is released from the recombinant *E. coli* by similar procedures. For systems such as recombinant mammalian cells, where much milder conditions are required to release the product, preparation of the antigen should be done in a similar manner. An alternative strategy, used particularly in the case of proteins

Method Table 1. Production of broad specificity antisera against *E.coli* K12.

1. Grow *E.coli* K12 to log phase and pellet the cells by centrifugation.
2. Suspend 0.32 g of *E.coli* pellet in 1 ml of lysis buffer (8.0 M urea, 2% Nonidet P40, 5% 2-mercaptoethanol).
3. Lyse by freezing and thawing 3-5 times.
4. To 200 μl of lysate add 300 μl of water and 1.5 ml of Freund's complete adjuvant. Emulsify by mixing in two syringes connected with a 19-gauge needle.
5. Inject the 2 ml of emulsion into 20-40 sites on the back of a New Zealand White rabbit. Each injection site should be intradermal.
6. Boost after 8 weeks, and at subsequent 3-week intervals by injection into single intramuscular sites of similar quantities of *E.coli* lysate emulsified in Freund's incomplete adjuvant.
7. 10 days after the fifth boost, bleed the animal out and prepare serum.

Method Table 2. Detection of *E.coli* proteins in SDS gels by immunoblotting[a].

1. Fractionate samples (200−500 μg) of the purified protein to be tested by SDS−PAGE[b]. The gel should also include samples (0.1−10 μg total protein) of the *E.coli* preparation.
2. Electrophoretically transfer the proteins from the running gel on to a nitrocellulose sheet[b].
3. Incubate the nitrocellulose sheet for 1 h in phosphate-buffered saline (8 mM Na_2HPO_4, 1.5 mM KH_2PO_4, 0.14 M NaCl) containing a suitable blocking protein (e.g. 3% bovine haemoglobin or 5% dried skimmed milk).
4. Incubate overnight at room temperature with gentle shaking in 50 ml blocking buffer containing 0.25 ml of rabbit anti-*E.coli* serum.
5. Wash the nitrocellulose sheet with at least 6 changes of blocking buffer; each wash should last 5−10 min.
6. Incubate for 4 h with radioiodinated anti-rabbit IgG (10^7 Bq in 20−50 ml).
7. Wash with at least 6 changes of PBS.
8. Dry the sheet and visualize bands by autoradiography for 1−3 days using Kodak X-S1 film in a cassette fitted with an intensifying screen.

[a]Examples of suppliers of equipment and reagents: Electroblot apparatus, Model EC420, Uniscience Ltd; nitrocellulose sheets, Anderman and Co. Ltd; intensifying screens, Cuthbert Andrews Ltd; affinity-purified antisera to rabbit IgG, Sigma Chemical Co. IgG fractions may be radiolabelled with ^{125}I using standard procedures (5).
[b]Suitable methods for SDS−PAGE and transfer to nitrocellulose are given in Chapter 1 of the companion volume (18).

produced by recombinant DNA systems, is to construct a host−vector system identical to the production system but lacking the gene insert coding for the product. This is then taken part of the way through the purification process, producing a set of proteins of similar biochemical properties to the product which is then used to raise antibodies.

3.2.2 *Immunochemical techniques*

A range of techniques is available in which antibodies are used to visualize contaminants after separation by various techniques such as gel diffusion, electrophoresis in agarose gels and electrophoresis in SDS-acrylamide gels. Perhaps the most useful of such methods are the many techniques of immuno-blotting (Western blotting). An example is given in *Method Table 2*.

Many refinements to the basic technique are employed to increase either the sensitivity or the convenience of the method. The two most commonly employed are affinity-purification of the antiserum, and the use of enzyme labelled (either alkaline phosphatase or peroxidase) rather than radiolabelled antibodies to visualize the antibody − antigen complex.

One useful method is the peroxidase/anti-peroxidase method using an antibody raised in rabbits against peroxidase. A soluble complex is allowed to form between this antibody and the peroxidase, in which the enzyme retains its activity. In the example given (*Method Table 3*), rabbit − anti-*E. coli* antibodies bound to the nitrocellulose membrane are visualized by allowing the donkey − anti-rabbit serum to form a bridge, by binding both to the rabbit antibodies on the nitrocellulose, and also to the rabbit − anti-peroxidase antibody. Peroxidase is detected by the conversion of the substrate diaminobenzidine to an insoluble form which precipitates around the band of *E. coli* protein (*Figure 1*).

3.2.3 *Immunoassay techniques*

Antibodies raised to partially purified host cell protein preparations which are considered likely contaminants have been used in radio-immunoassays (RIA) in which the displacement of radiolabelled host cell protein from the antibody by cold (unlabelled)

Method Table 3. Peroxidase − anti-peroxidase (PAP) detection of bound rabbit antibodies[a].

---

1. Carry out SDS − PAGE and electroblotting as previously described.
2. Block the nitrocellulose by incubation in a solution of 5% powdered milk in phosphate buffered saline (PBS-milk) for 1 h.
3. Incubate overnight with gentle shaking in a 1/250 dilution of rabbit anti-*E. coli* antiserum in PBS-milk.
4. Wash with at least 6 changes of PBS-milk.
5. Incubate for 2 h in a 1/500 dilution of donkey anti-rabbit antiserum in PBS-milk.
6. Wash with at least 6 changes of PBS.
7. Incubate for 1 h in a 1/250 dilution of PAP complex in PBS.
8. Wash with at least 6 changes of PBS.
9. Incubate with 25 − 50 ml of freshly made substrate (60 mg diaminobenzidine, 30 μl 30% hydrogen peroxide, 100 ml water).
10. Bands develop in 5 − 20 minutes and are seen as dark red on a pale red background.

---

[a]A soluble complex of rabbit anti-peroxidase and horseradish peroxidase is widely available (e.g. Sigma Chemical Co.). Most preparations of powdered milk are suitable for blocking nitrocellulose groups.

A.F.Bristow

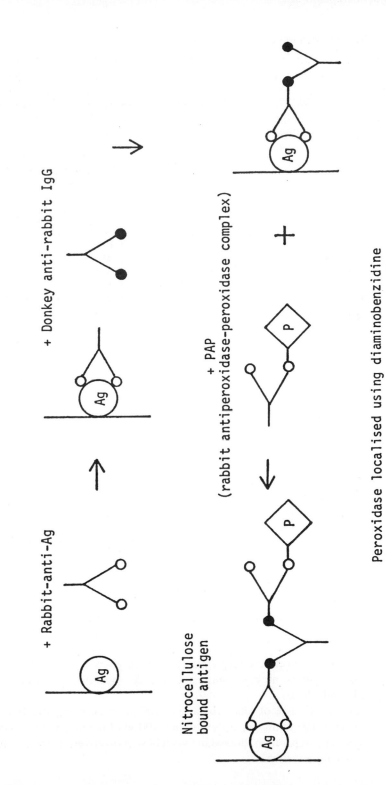

Peroxidase localised using diaminobenzidine

Figure 1. Peroxidase–anti-peroxidase (PAP) staining.

35

Method Table 4. Radioimmunoassay for *E. coli* proteins.

A. *Reagents*

1.   *E. coli* proteins: having prepared a crude lysate of *E. coli*, as previously described, it is then necessary to prepare a partially purified preparation in a way that utilizes similar techniques to those used in the preparation of the product of interest. In this way the proteins detected in the immunoassay will be biochemically similar to the product and represent likely contaminants. Perhaps the most useful approach is to prepare a Sephadex fraction equivalent in molecular weight to the product.

2.   Radioiodination: a number of methods are suitable and are described in (2). The simplest approach is the chloramine-T method. Briefly, incubate 10 μg of *E. coli* protein with 1 mCi of [^{125}I]sodium iodide in a volume of 30 μl in 0.4 M phosphate buffer pH 7.0. Add chloramine-T (10 μl of 1 mg ml^{-1} solution in a phosphate buffer) and allow the reaction to continue for 30 sec before stopping with sodium metabisulphate (5 mg ml^{-1} in phosphate buffer). Separate the radioiodinated protein preparation (tracer) from the free iodine by gel filtration on an appropriate grade of Sephadex in 0.1 M phosphate buffer containing $0.1 - 1.0$ mg ml^{-1} bovine serum albumin.

3.   Assay buffer: the radioimmunoassay should be carried out in phosphate-buffered saline containing 0.5% bovine serum albumin.

B. *Method*

1.   Antiserum titration curves:

    (i)   Set up a range of tubes containing 100 μl assay buffer, 100 μl tracer (20 000 − 50 000 cpm in assay buffer) and either 100 μl of anti-*E. coli* antiserum, diluted between 1/500 and 1/500 000, or 100 μl assay buffer (non-specific binding, NSB).

    (ii)   Incubate overnight at 4°C.

    (iii)   Precipitate the antibody bound fraction in the following way: add 300 μl of PEG/gamma globulin reagent (5 g polyethylene glycol 6000, 30 mg bovine gamma globulin, 20 ml 0.05 M Tris−HCl pH 8.5).

    (iv)   Incubate for 1 h at 4°C.

    (v)   Centrifuge at 2500 rpm for 30 min, aspirate off the supernatant and count the pellets.

2.   Assay:

    (i)   Select a concentration of antiserum which gives a specific binding of $15-40\%$ at zero antigen dose, B_0 (specific binding = % bound − NSB).

    (ii)   Set up a range of tubes containing 100 μl of assay buffer, 100 μl tracer, 100 μl diluted antiserum and 100 μl of the unlabelled *E. coli* protein preparation, diluted in two-fold steps to cover the range 1 pg to 1 μg per tube.

(iii) Carry out the incubation and separation steps as described. The working range of the assay is the range of concentrations of the *E.coli* protein preparation which result in the complete displacement of the antibody-bound tracer, from the level at zero *E.coli* protein concentration down to the non-specific binding level.

(iv) To determine levels of contamination include tubes containing product in the assay, and read off the displacement curve.

Method Table 5. Ultrasensitive silver stain.

---

1. Carry out SDS-gel electrophoresis of test proteins and *E.coli* lysates as previously described.
2. Being careful to avoid touching the gel, transfer it to 50% methanol, 10% acetic acid, and leave overnight to fix.
3. Wash, with gentle shaking at 60°C in 3 changes of 5% methanol, 7% acetic acid, for 10, 20 and 30 min.
4. Incubate for 10 min at 60°C in a 1/10 dilution of commercial $37-40\%$ formaldehyde[a], adjusted to pH 7 with sodium hydroxide.
5. Wash at 60°C with 3 changes of distilled water at pH 7, for 10, 20 and 30 min.
6. Equilibrate the gel in water at room temperature.
7. Prepare the silver reagent as follows: add 4 ml commercial $25-30\%$ methylamine to 20 ml 0.36% sodium hydroxide. Add the resulting solution to 8 ml 20% silver nitrate in water, dropwise with mixing, until the brown precipitate just clears. Filter and adjust to 200 ml with water.
8. Incubate the gel in the silver reagent for 10 min at room temperature.
9. Wash briefly with distilled water and then incubate in 0.005% citric acid, 0.019% formaldehyde. This solution should be changed at 10-min intervals, or whenever it starts to discolour.
10. Protein bands will appear in 10-30 min.
11. Stop the staining procedure when the bands stop increasing in intensity relative to the background, by transferring to distilled water.

---

[a]Formaldehyde is toxic. Step 4 should be carried out in a fume hood. Silver nitrate is corrosive and should be handled with care.

host-cell protein is used to obtain a quantitative measure of host-cell protein present. The method given may be refined in a number of ways, and the reader is referred to a general text on immunoassays (e.g. ref. 5) when setting up this type of procedure. An example is given in *Method Table 4*.

3.3 Ultrasensitive staining techniques

A number of techniques for detecting proteins in gels have now been developed in which the sensitivity afforded by immunochemical methods is approached. The most widely used of these ultrasensitive staining techniques is the silver stain, which exists in many different forms. The method described (*Method Table 5*) is that of Marshall (6). More

37

recently, techniques based on colloidal gold have been developed which are claimed to offer even greater sensitivity. The AuroDye reagents (Janssen Life Science products) offer an easy and extremely sensitive method for detecting proteins after transfer to nitrocellulose membranes.

3.4 **Quantitation and sensitivity**

Quantitation of these ultrasensitive protein detection methods, particularly the immunochemical and photochemical gel techniques, presents a number of difficulties. Although band intensities on both stained gels and autoradiograms are easy to assess by densitometry, such measurements are not absolute and require comparison with standards. The sensitivities of both silver-staining and immunochemical staining vary greatly from protein to protein, however, and results obtained by comparison with unlike standards may be unreliable. Also, when using these techniques at the limit of their sensitivity, slight variation in experimental procedure or in reagents may seriously affect the results. Quantitative data are easier to obtain using immunoassays, although a number of factors can cause artefactual results in such techniques. For instance, it is often difficult to standardize a mixture of antigens. Ideally, a single standard preparation should be made and retained. Also, many proteins are damaged during radioiodination, leading the antibody to discriminate between the labelled and unlabelled antigen.

Notwithstanding the problems of quantitation, approximate limits of sensitivity might be 5 p.p.m. for immunoassay, 25−50 p.p.m. for immunoblotting and 200−400 p.p.m. for silver-staining. Nevertheless, each technique offers unique advantages; the immunoassay offers ultimate sensitivity, the immunoblotting approach gives qualitative as well as quantitative information on antigens present, and the silver-staining approach does not rely on the production of antibodies in experimental animals in order to detect a contaminant.

3.5 **Nucleic acid hybridization techniques**

Techniques for the detection of picogram quantities of DNA for the most part depend upon hybridization with radiolabelled DNA probes derived from either the host cell or the recombinant vector used to produce the product. The methods given in *Method Table 6* are the subject of many variations and refinements, and the reader is referred to an earlier volume in the series covering the theory and practice of nucleic acid hybridization (7).

3.6 **The *Limulus* test for bacterial endotoxin**

Pyrogenic fragments of bacterial cell walls may be detected either by elevation of temperature in the rabbit or by their ability to cause gelation of a lysate of amoebocytes derived from the blood of the horseshoe crab, *Limulus*. The LAL test (*Limulus* amoebocyte lysate) is now widely used in quality control to check for the presence of bacterial pyrogens in medicines that are given parenterally, and is particularly appropriate for the analysis of proteins produced by fermentation in Gram-negative organisms such as *E.coli* (*see Method Table 7*).

In more recent variations of the *Limulus* test, the formation of the clotting enzymes in the lysate may be determined using a chromogenic substrate (9). This procedure,

Method Table 6. Dot-blot hybridization analysis of *E.coli* DNA.

A. *Reagents*

1. *E.coli* DNA: total chromosomal DNA from *E.coli* may be readily extracted and purified by standard techniques involving lysis of the cell pellet, extraction of the protein with phenol, and back-extraction with chloroform to remove phenol.
2. Radiolabelling: total *E.coli* DNA is radiolabelled to high specific activity with ^{32}P-nucleotides by the process of nick-translation. For performing a limited number of analyses it is easier to purchase the nick-translation reagents in 'kit' form (Amersham International).
3. SSC is standard saline citrate buffer. It is prepared as a 20 × concentrate. Dissolve 175.3 g of NaCl and 88.2 g sodium citrate in 800 ml water. Adjust the pH to 7.0 with sodium hydroxide and make up to 1 litre.

   Detailed procedures for the extraction and manipulation of DNA may be found in a number of laboratory manuals to which the reader is referred (e.g. ref. 8).

B. *Method*

1. Soak a nitrocellulose membrane in 2× SSC.
2. Apply the reference (unlabelled) *E.coli* DNA preparation at concentrations over the range 0.1 pg ml^{-1} to 100 ng ml^{-1} as a series of dots to the nitrocellulose filter. The use of a commercial dotting apparatus (e.g. Bio-Rad) will allow volumes up to 150 μl to be applied. In another range of dots apply the test protein at concentrations between 1 μg ml^{-1} and 10 mg ml^{-1}.
3. Dry the filter in air and bake at 80°C for 1 h to fix the dots.
4. Prehybridize the filter by incubating it overnight at 50°C with gentle shaking, in a sealable plastic bag with a minimum volume (7 ml for a 5 × 5 cm filter) of prehybridization solution (1% Ficoll, 1% polyvinylpyrollidone, 1% bovine serum albumin, 5 × SSC, 0.5 ng ml^{-1} salmon sperm DNA).
5. Remove the prehybridization solution from the bag and replace with the same volume of prehybridization solution containing 10^7 cpm of radiolabelled *E.coli* DNA. Incubate overnight as before.
6. Remove the filter from the bag. Wash at 50°C in 2 × SSC, 0.1% SDS, 3 changes.
7. Dry the filter in air and determine the bound radioactivity by autoradiography. The content of DNA in the sample may be assessed by comparison with the test dilution series.

---

performed in microtitre well plates, offers the advantages of economy and improved quantitation.

Some products or reagents, such as chelating ligands and certain vaccine preparations, interfere in the *Limulus* test (for a complete list see ref. 2), and Gram-positive endotoxins do not work in the *Limulus* test. It may be necessary therefore to demonstrate absence of pyrogens using the rabbit test. In this procedure, pyrogenic substances injected into

Method Table 7. The *Limulus amoebocyte* lysate test (LAL) for bacterial endotoxins.

A. *Reagents*

1. Lysate: *Limulus amoebocyte* lysate is available from a variety of sources and is of variable sensitivity. Lysate produced by Atlas Bioscan Ltd; may be used in the method being described here.
2. Standard: use a working standard of endotoxin that has been calibrated in IU ml^{-1} against the International Standard of Endotoxin for LAL gelation tests (84/650 obtainable from the National Institute for Biological Standards and Control).
3. Laboratory ware: all pipette-tips, containers and reagents must be purchased as 'sterile and pyrogen-free', otherwise they must be made pyrogen-free using procedures described in Section 4.4.

B. *Method*

1. Carry out the experiment under clean conditions at 37°C. A laminar flow hood containing a thermostatted heater is suitable, although strictly aseptic conditions are not essential. Place an opened sterilin petri dish on top of a petri dish lid upon which has been drawn a grid dividing the dish up into 16 numbered sectors.
2. Pipette a 10 μl droplet into each sector.
3. Add 10 μl of either endotoxin standard or test at the dilutions shown to each droplet.

Grid No.	Addition	Concentration
1	Endotoxin standard	8 IU ml^{-1}
2	Endotoxin standard	4 IU ml^{-1}
3	Endotoxin standard	2 IU ml^{-1}
4	Endotoxin standard	1 IU ml^{-1}
5	Endotoxin standard	0.5 IU ml^{-1}
6	Endotoxin standard	0.25 IU ml^{-1}
7	Endotoxin standard	0.125 IU ml^{-1}
8	Endotoxin standard	0.063 IU ml^{-1}
9	Endotoxin standard	0.031 IU ml^{-1}
10	Endotoxin standard	0.015 IU ml^{-1}
11	Pyrogen-free water	
12	Test solution	undiluted
13	Test solution	1:10
14	Test solution	1:100
15	Test solution	1:1000

4. After 1 h add by microsyringe 1 drop of methylene blue solution (a 2% solution in water, diluted 1:10 in ethanol before use) to each droplet.
5. A positive result is indicated by the presence of a blue star on the surface of the droplet. A negative is indicated by a loss of the surface tension and the formation of a completely blue spot. The pyrogen-free water control must be negative for the test to be valid.

6.   If standards 1 − 8 are positive, and 9 and 10 negative, the sensitivity of the lysate is defined as 0.06 IU ml^{-1}. If test solutions 12 and 13 are positive and 14 and 15 negative, the content of endotoxin in the text solution is more than 0.6 IU ml^{-1} but less than 6.0 IU ml^{-1}. The concentration may then be more closely bracketed by repeating the assay with two-fold dilutions of test solution.

---

rabbits produce an elevated temperature. The assay of pyrogens in rabbits is an exacting technique, requiring careful animal husbandry and experienced operators, and is outside the scope of this book. The interested reader is referred to Pearson (2).

3.7 Detection of viruses

3.7.1 *Human viruses*

For proteins derived from purified human tissue or serum, the viruses of greatest concern are the AIDS virus (HIV) and hepatitis B. Both viruses may be detected in protein preparations using immunological tests which detect the presence of viral antigens. These reagents are available in kit form from a number of manufacturers, typical examples being the Wellcozyme test for HIV antibody, and Hepatest, for hepatitis antigen, both marketed by Wellcome Diagnostics.

3.7.2 *Murine viruses*

More than 20 viruses of murine origin have been identified as possible contaminants in monoclonal antibodies derived from murine hybridomas. Of these, the Hanta virus (haemorrhagic fever) and lymphocytic choriomeningitis virus have the potential to cause serious disease in man, whilst a number of viruses may be of concern in the immune-deficient patient. The most useful method of detecting specific murine viruses in preparations is the mouse antibody production test (MAP). Preparations to be tested are injected into healthy mice, and after suitable periods of time the mouse serum is checked for the presence of antibodies to the virus of interest. Detailed descriptions of such methodology are outside the scope of this book, since the tests can only be performed in fully established virology laboratories. Nonetheless, such testing is an important aspect of verifying the appropriate purity of monoclonal antibodies of therapeutic use.

4. PROCESS VALIDATION

Since we have identified a number of possible contaminants as being of potential concern, it is desirable to design the purification process such that they are specifically excluded from the product.

4.1 DNA

Generally speaking, it is not necessary to include steps specifically to remove DNA. It is a strongly acidic molecule and is usually effectively removed by any ion-exchange step in the process. Other chromatographic or bulk separation steps usually reduce the DNA content by one or two orders of magnitude also.

4.2 **Proteins**

Proteins are probably the hardest class of contaminant to separate from the product, since those remaining with the product by the end of the purification may be of similar molecular weight, isoelectric point and hydrophobicity. Apart from the reintroduction of the affinity purification step at a later stage in the process, there are not really any specific measures to be taken to remove proteins of similar characteristics to the intended product and they are usually reduced to the desired level by the inclusion of extra purification steps such as the use of both cation- and anion-exchange chromatography.

4.3 **Aggregated or degraded products**

Forms of the desired protein that have suffered some form of degradation represent a particular class of protein contaminants.

A number of protein modifications are known to occur during purification processes. These include:

(i)   dimerization or aggregation which may be covalent or non-covalent;
(ii)   oxidation, usually of methionine residues;
(iii)   deamidation of asparagine and glutamine residues;
(iv)   incorrect folding;
(v)   incorrect disulphide formation;
(vi) β-aspartyl shift, in which a β peptide bond is formed through the β-carboxyl group of aspartate residues;
(vii)   endo- or exopeptidase cleavage.

Oxidized and deamidated forms are charge-variants and may be removed by ion-exchange procedures. Aggregated forms can be removed by size-exclusion chromatography. Wrongly disulphide bonded forms would usually have quite different properties and be easily removed. Other forms such as wrongly folded forms, β-aspartyl shifts and forms partially degraded at the termini may be more difficult. Preparative HPLC, if applicable to the product, is often useful. Otherwise, low levels of contamination with such forms may be inevitable, and the onus would be on clinical studies to demonstrate safety.

4.4 **Pyrogens**

Bacterial endotoxins are lipopolysaccharides and as such have physicochemical properties quite different from most proteins. However, although protein purification techniques may reduce pyrogen level, its complete removal may necessitate specific depyrogenation steps. The properties of endotoxins are summarized in (2). A number of these properties have been used to design depyrogenation procedures. Monomeric lipopolysaccharide (LPS) has a molecular weight of 10 000 to 20 000. It readily aggregates under aqueous conditions, however, to form aggregates of up to 1 million daltons. Ultrafiltration through a 100 000 molecular weight cut-off membrane is therefore a particularly effective step (10). Conversely, for high-molecular-weight products the endotoxin may be disaggregated with chelating agents and passed through a 50 000 molecular weight cut-off membrane while the product is retained (11). Endotoxin is negatively charged and this has allowed the use of positively charged membranes such as polyamides or bonded amines for their removal (12). Aliphatic polymers such as polypropylene bind endotoxin and have also been used to depyrogenate pharmaceutical preparations (13). It has also been reported that histamine binds endotoxin,and histamine coupled to Sepharose is being investigated as a possible depyrogenating matrix for proteins (14). It should be noted, however, that care must be taken to avoid reintroducing pyrogenic material during the process. All reagents used must be sterile and pyrogen-free, the simplest procedure being autoclaving (120 psi, 20 min, 3 cycles). Glassware must be dry-sterilized at 240°C for 1 h. Laboratory plasticware such as tubing may be depyrogenated by treatment with either acid or alkali (2). All procedures must be carried out under aseptic conditions. Many reagents, for example antibody affinity columns may not be treated in this way. Under these circumstances the best procedure is extensive washing with pyrogen-free buffers until the eluate is pyrogen-free, as determined by the LAL test.

4.5 Viruses

As with pyrogens and DNA, virus particles are biochemically different from most proteins of therapeutic interest, and inclusion of specific steps for their removal is often not necessary. However, particularly in the case of proteins purified from human tissue or human cell lines that may be contaminated with pathogenic human viruses, it may be considered necessary to take specific steps to inactivate viruses. Heat treatment (60°C, 10 h) has been widely used to inactivate viruses in serum proteins. Other procedures employed have been UV irradiation (15), ionizing radiation (16) and chemical modification with formaldehyde or β-propionolactone (17). The exact conditions employed must be determined for specific virus, however, since viruses vary widely in their sensitivity to each type of inactivation. Moreover, although all three procedures successfully inactivate viruses, they are potentially capable of causing denaturation or chemical modification of proteins and should be applied with caution.

4.6 Leakage products from matrices

Affinity matrices (such as antibody − Sepharose conjugates) may be unstable over a long period of time. As a result, the conjugated ligand may 'leak' from the column and contaminate the product. Alternatively, following repeated usage, affinity columns may become fouled with impurities which eventually start appearing in the eluate. For these reasons it is usually desirable that affinity purification should not be the final stage in a purification process, and subsequent gel-filtration or ion-exchange steps may be required.

4.7 Clearance studies

For contaminants such as viruses or DNA, lower limits of contamination may be difficult to set, since in principle at least, contamination with a single virus particle may be of clinical consequence. Since the actual level of contamination is usually well below the lower limit of detection of the analytical techniques available, it is usually desirable to predict the actual level of contamination by performing clearance studies, in which the effectiveness of each step in a process at removing a given contaminant is measured by 'spiking' with preparations of the contaminant. The overall purification factor for the entire process can then be calculated (*Table 1*).

In the example given, DNA has been determined by hybridization analysis. This kind

Table 1. Process validation[a]: DNA clearance studies.

Purification step	DNA spike	DNA content after step	Clearance factor
1. Diafiltration/ concentration	$10 \ \mu g \ ml^{-1}$	$0.2 \ \mu g \ ml^{-1}$	5×10
2. Affinity chromatography on antibody-coupled matrix	$10 \ \mu g \ ml^{-1}$	$0.05 \ \mu g \ ml^{-1}$	2×10^2
3. Anion-exchange chromatography	$10 \ \mu g \ ml^{-1}$	$0.02 \ ng \ ml^{-1}$	5×10^5
4. Gel-filtration	$10 \ \mu g \ ml^{-1}$	$0.025 \ \mu g \ ml^{-1}$	4×10^2
		Overall process clearance factor	2×10^{12}

[a]The hypothetical purification process involves four purification steps. Before each step the partially purified preparation is 'spiked' with a known concentration of DNA. The DNA remaining after each step is determined by dot-blot hybridization analysis. Clearance factors for each step are calculated. The clearance factor for the whole process is the product of those for each individual step.

of study may equally well be performed using ^{32}P-labelled DNA, often with increased precision. The DNA used for spiking in these studies should be that which the process is intended to remove; for example, if the recombinant DNA expression system is a Chinese Hamster ovary (CHO) cell, then CHO cell DNA should be used. In the case of virus clearance studies, the viruses used should be identical, or related, to potential contaminants in the product. Thus in the case of murine hybridomas, murine retroviruses would be appropriate, whilst for human serum proteins, hepatitis B or even HIV clearance studies would be necessary.

5. REFERENCES

1. Lloyd,G. and Jones,N. (1986) *J. Infect.*, **12**, 117.
2. Pearson,F.C. (1985) *Pyrogens: Endotoxins, LAL Testing and Depyrogenation.* Marcel Dekker, New York.
3. Bloom,S.R., Barnes,A.J., Adrian,T.E. and Polak,J.M. (1979) *Lancet,* i, 14.
4. Johnstone,A. and Thorpe,R. (1982) *Immunochemistry in Practice.* Blackwell Scientific, Oxford.,
5. Thorell,J.I. and Larson,S.M. (1978) *Radioimmunoassay and Related Techniques.* C.V.Mosby, St Louis.
6. Marshall,T. (1984) *Anal. Biochem.*, **136**, 340.
7. Hames,B.D. and Higgins,S.J. (eds) (1985) *Nucleic Acid Hybridization: A Practical Approach.* IRL Press, Oxford.
8. Maniatis,T., Fritsh,E.F. and Sambrook,J. (1982) *Molecular Cloning.* Cold Spring Harbor Laboratory, Cold Spring Harbor, New York.
9. Harris,R.I., Stone,R.C.W. and Stuart,J. (1983) *J. Clin. Path.*, **36**, 1145.
10. Nelson,L.L. (1978) *Pharm. Technol.*, **2**, 46.
11. Sweadner,K.S., Forte,M. and Nelson,L.L. (1977) *Appl. Environ. Microbiol.*, **34**, 582.
12. Fiore,J.V., Olson,W.P. and Holst,S.L. (1980) In *Methods of Plasma Protein Fractionation.* Curling,J. (ed.), Academic Press, London.
13. Robinson,J.R., O'Dell,M.C., Tabacs,J., Burnes,T. and Genovesi,C. FDA monograph on depyrogenation.
14. Minobe,S., Sabo,T., Tosa,T. and Chibata,I. (1983) *J. Chromatogr.*, **262**, 193.
15. Kleczkowski,A. (1968) In *Methods In Virology.* Maramorosch,K. and Koprowski,H. (eds), Academic Press, New York, Vol. 4, p. 93.
16. Ginoza,W. (1968) In *Methods in Virology.* Maramorosch,K. and Koprowski,H. (eds), Academic Press, New York, Vol..4, p. 139.
17. Potash,L. (1968) In *Methods in Virology.* Maramorosch,K. and Koprowski,H. (eds), Academic Press, New York, Vol. 4, p. 371.
18. Harris,E.L.V. and Angal,S. (eds) (1989) *Protein Purification Methods: A Practical Approach.* IRL Press, Oxford.

CHAPTER 3

Purification for crystallography

S.P. WOOD

1. INTRODUCTION

In the early days of protein biochemistry, crystallization was often employed as a purification technique and crystallinity was considered an index of purity. Crystallographic studies were focused on a selection of proteins whose principal qualification was their abundance and ease of crystallization. The techniques of protein crystallography have now developed to such a degree that a complete three-dimensional structure analysis may not be a long task. Indeed, it is often the production of suitable crystals that is rate-limiting for the study of many proteins. The following observations serve to set the problem. Even for those proteins which can be crystallized from very impure mixtures and thus purified, crystal quality improves with each recrystallization as purity increases. Failure to consistently grow crystals of high quality afflicts many structure analyses. In view of the simplicity of crystal-growing techniques and the relative complexity of purification procedures, it seems likely that the variability derives from the latter, implying the importance of purity. However, it is a common observation that apparently identical crystallizations made from the same batch of protein can yield crystals of variable quality and often no crystals at all! Homogeneity of the protein is clearly only one of the important factors in crystal growth, but we are well equipped with methods to try to control this parameter. More detailed discussion of the theory and practice of crystal growth can be found in refs 1−4. The relationship of homogeneity and crystal growth is not well documented in the literature. Reports of successful crystal growth are numerous, but many of the problems encountered are not fully recorded. This section describes the nature of protein crystals and how they are formed in order to examine the importance of homogeneity.

The protein crystal is a rather open three-dimensional lattice, where each repeating motif is a single protein molecule or group of molecules. Much of the crystal volume is occupied by water molecules (30−80%) and only a small portion of the protein (or aggregate) surface is involved in contacts with other protein molecules. Those water molecules close to the protein surface are well organized in hydrogen bonding networks with the protein and with each other, while further out, in the solvent channels of the crystal, their properties are more like those of bulk water. The types of interactions involved in contacts between protein molecules are very much like those believed to stabilize protein structure, namely, multiple weak forces involving hydrogen bonding, ion-pairs, hydrophobic interactions and occasionally metal coordination. As a result the crystals are soft and extremely sensitive to environmental factors such as humidity. The integrity of the crystal is dependent on the almost faultless repetition of these protein−protein contacts involving perhaps 10^{16} molecules. Crystals of small molecules in general contain no free solvent, and intermolecular interactions are far more extensive. The crystals are hard, and the crystallization process is an extremely powerful purification procedure widely used in all branches of preparative chemistry.

The basic principles of protein crystallization are the same in most respects as those which form the foundation for the more familiar techniques of crystallizing small molecules like salts or amino acids. It is necessary to achieve a slow approach to a low degree of supersaturation of protein in solution, that is, a condition in which the solvent holds more protein in solution than is normally possible for a true equilibrium of minimum free energy. This metastable protein solution might be achieved by a gradual change of precipitant or protein concentration, by a change of pH or temperature. The final conditions should be such that the crystalline state is thermodynamically most stable. During searches for appropriate conditions, amorphous states are achievable from a sometimes distressingly wide variety of conditions, reflecting their probably random bonding network. The crystalline state, when achieved, is usually of far lower solubility than the closest amorphous state, and sometimes crystal growth is sustained at the expense of redissolving of coexisting amorphous material. Techniques of dialysis, vapour diffusion and slow cooling or warming are used to approach the required conditions. During protein purification, valuable information may be gained about the efficacy of different protein precipitants and about the stability of the protein. Limits of protein solubility with pH and estimates of isoelectric point can be determined. Changes in apparent molecular weight with conditions might provide clues to specific aggregation processes necessary for crystal packing. Any factor which influences protein solubility might be manipulated to encourage a marginal dominance of the appropriate protein− protein interactions. Most proteins exhibit a solubility minimum at their pI where net charge is zero and repulsive forces are minimal. Many crystals grow close to these conditions. However, proteins are very complex polyions with particular patterns of charge on their surface which give rise to multiple solubility minima. This is reflected in the growth of many protein crystals several pH units away from the pI. In such conditions of low solubility, appropriate molecular aggregates slowly reach a critical size or nucleus from which the growth of the macroscopic crystal proceeds as a spontaneous process. The slow approach and low degree of supersaturation are important since the slow diffusion rates of macromolecules will limit the rate of presentation of appropriately oriented molecules to growth points and the number of such 'nuclei' is related to supersaturation. The degree of supersaturation required to initiate homogeneous nucleation (that is, nuclei generated from protein aggregates) is considered to be much greater than that required to sustain growth. This explains why it is often possible to produce showers of small crystals from excess nucleation, but much more difficult to produce a small number of larger crystals. Fortunately, at low degrees of supersaturation nucleation often appears to be heterogeneous, with growth initiating on particles of dust, fibre or hair which in spite of the experimenter's efforts always seem to gain entry to the crystallization.

2. METHODS FOR PROTEIN CRYSTALLIZATION

Crystal-growing techniques are numerous, with particular favourites existing in laboratories where they have been successful. A comprehensive coverage of these methods is available elsewhere (1−4). Here I present a brief description of some of the more popular and successful methods which offer a reasonable starting point for anyone about to set out to crystallize their protein.

2.1 **Equipment**

The equipment requirements are not extensive. The most expensive items are good microscopes (ideally with photographic facilities). Stereo-zoom microscopes with polarizer/analyser are rather useful for examination of crystal-growing containers, where it may be necessary to try to focus on crystals through variable thickness of glass, plastic and solution, as well as for manipulating larger crystals prior to investigation by X-rays. Polarized light is very useful in locating optical anisotropy prior to X-ray work. A conventional or inverted microscope must be used to identify smaller crystals. Again, polarized light is particularly useful for locating microcrystals amongst amorphous deposits. The depth of focus is of course very limited and the image easily degraded by extraneous materials in the light path. A selection of pipette guns and syringes is required for dispensing small volumes of liquid, and a high-speed (15K rpm) microfuge is ideal for removing large particulates from protein solutions. Centrifugal ultrafiltration devices are useful for concentrating and performing rapid buffer changes on small quantities of protein.

Most of the apparatus in which crystals have been grown has not been specialized, but rather borrowed from other techniques or merely assembled from standard laboratory glass and plasticware. The nature of the apparatus is dependent on the abundance of the protein and the conditions that are to be used to induce crystal growth. Many crystals have been grown by changing the ionic strength, alcohol content or pH of a protein solution. This is done slowly, using dialysis or vapour diffusion, which is described below.

2.2 **Methods for growing crystals**

2.2.1 *Vapour diffusion*

As an example let us assume that 10 mg of protein is available and little is known about its solubility or stability. The protein has been concentrated to 0.5 ml in a relatively low ionic strength buffer, say 2 mM Tris−HCl at pH 8, with no problems. It is advisable to start out with the highest protein concentration that will permit a reasonably wide survey of conditions with the material available. A reduction of the protein concentration may be useful later on as a variable to improve crystal size. Many proteins are not soluble in water, and it is wise therefore to maintain some buffer salts, but these should be sufficiently dilute to permit change in pH by addition of small volumes of other more concentrated buffers. Inclusion of NaCl or KCl may be required to 'salt in' some proteins.

Initial screening experiments are set up according to the following method:

(i)   Remove particulates by centrifugation and pipette 10 μl of the protein solution on to a clean silanized glass coverslip.

(ii)   Adjust pH by the addition of 1 μl of a concentrated ($\sim$1M) buffer of the desired pH and composition.

(iii)   Increase the drop volume to 20 μl by the addition of a precipitant (e.g. 15%

w/v polyethylene glycol, PEG). Drop volumes greater than 20 μl should be avoided.

(iv)   Invert the coverslip and suspend the 'hanging drop' over a reservoir of the precipitant (e.g. 15% PEG). The seal between the coverslip and the precipitant reservoir should be achieved with a layer of silicone vacuum grease.

(v)   Incubate the apparatus (e.g. at 22°C). Water will evaporate from the drop, allowing both protein and precipitant concentration to increase, thereby producing a supersaturated solution from which crystals may grow. The rate of equilibration is temperature-dependent, and crystals may take from several hours to several months to grow.

A comprehensive screen of conditions requires large quantities of protein ($\sim$ 100 mg) and is labour-intensive. With limited resources, it is best to concentrate on a small number of parameters known to be crucial: (i) *pH*: 0.5 pH unit intervals are suitable to begin with; (ii) *temperature*: 4°, 22° and 37°C are most commonly useful; and (iii) *precipitants*: those that have been particularly useful include ammonium sulphate, sodium chloride, polyethylene glycol, ethanol and 2-methane-2,4-pentanediol (MPD) as well as low ionic strength.

Arrays of 'hanging-drop' experiments should be set up on the initial screen. Plastic tissue culture trays (e.g. 24-well plates) are most convenient, *Figure 1*. If limits of solubility or other special requirements are known from studies during purification, the size of the screen can be dramatically reduced. The appearance of microcrystals or a sharp solubility change indicates where resolution of the screen can be increased. Finer intervals of pH, temperature or less severe precipitant strength can be used to refine conditions for crystal growth. If the initial screen fails to produce crystals, the most hopeful conditions should be repeated, with inclusion of the following possible additives:

(i)   *metal ions*: Cu^{2+}, Zn^{2+}, Ca^{2+}, Co^{2+};

(ii)   *chelators*: EDTA;

(iii) *low-molecular-weight organic compounds*: acetone, dioxane, phenol;

(iv)   *detergents*: n-octyl-β-D-glucopyranoside, up to the critical micelle concentration;

(v)   *ligands*: substrates, cofactors, other modifiers.

Smaller initial drop sizes can be used and, indeed, are recommended when using ethanol or detergent, but the procedure should be modified to permit reliable control of drop composition, particularly of pH. Adjustment of pH can be carried out on a larger volume followed by subdivision into smaller drops (e.g. 5 μl) allowing a larger number of additives to be tested. One of the most serious limitations of the 'hanging-drop' technique is the relatively large area of protein solution$-$air interface where some proteins form a 'skin'. If crystals do form and contact this sticky film, subsequent manipulations can become difficult. The effect can be reduced by keeping the equilibration interval to about 5$-$10% of the final concentration.

The above equilibration procedure of vapour diffusion can be carried out in many other ways. For instance, protein drops sitting in slide cavities or multiwell spot plates can be sealed in plastic sandwich boxes containing a precipitant reservoir, or small test-tubes containing protein can be allowed to stand in precipitant within a closed vial. Changes in pH can be accomplished by equilibrating with a volatile acid or base.

A

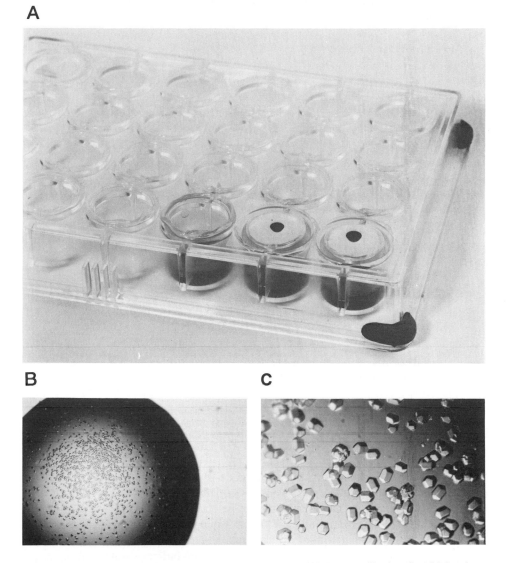

B   C

Figure 1 (a). The experimental set-up for a hanging drop vapour diffusion crystallization (6). A Linbro tissue culture tray (Flow Laboratories) provides 24 wells for 0.5 or 1 ml of precipitant reservoir. Protein drops hang from a coverslip which is sealed to the well rim with silicone vacuum grease. A small spacer should always be employed to ensure that the plate lid does not contact any excess grease around the coverslips lest removal of the lid for observation brings off all the coverslips with it. Coloured solutions have been used for the well and drop contents to aid inspection. (**b**), (**c**): small crystals (< 100 μm) of the aspartic proteinase of avian myeloblastosis virus produced in such apparatus at low and higher magnification to show what one might expect to see at a successful initial screen point. (Courtesy of J.Cooper and P.Stropp.)

2.2.2 *Dialysis*

Dialysis procedures have also been quite successful. Semipermeable membranes of the types commonly employed in preparative procedures are used to contain protein solution

Figure 2. A selection of apparatus suitable for dialysis crystallization experiments at various scales. **Left**: a microdialysis button is machined from a block of perspex to provide a protein chamber as small as 5-10 μl (Cambridge Repetition Engineers). **Centre**: this arrangement uses a thick-walled capillary as described by Zeppezauer (5). The dialysis membrane is held in place by a rubber gaiter. Charging this apparatus with protein solution by microsyringe or pasteur pipette requires some caution to avoid air bubbles and puncturing the membrane. The lower large gaiter should be slotted to avoid trapping air next to the membrane. **Right**: a small rimless test-tube (5 × 60 mm) is held by a plastic bottle cap (suitably holed with a hot test-tube) and closed with dialysis membrane. Volumes from 150 μl to 1 ml can be used.

in thick-walled capillary tubes, small test-tubes or perspex microdialysis buttons (see *Figure 2* for illustration). The membrane can be secured with an 'O' ring or gaiter cut from a suitable diameter of silicone rubber tubing. Polyacrylamide plugs can be employed to retain protein within capillary tubes. The device is then suspended or immersed in a dialysable precipitant, a low-ionic-strength buffer or water. The exterior conditions can be varied in a stepwise manner, with an equilibration interval of a few days between steps. Gradients of precipitant strength in the column of protein solution may lead to crystal growth at a particular height and aid choice of the ideal precipitant strength. In general, dialysis methods are not so economical on protein. Perspex buttons with an internal volume of 10 μl are available, but more than this is usually required to successfully fill a button without trapping air bubbles under the membrane. However, dialysis methods do allow more flexibility, as amorphous precipitates in failed attempts to produce crystals can be redissolved with appropriate solutions placed outside of the chamber, and another condition investigated with the same sample. Inverted test-tube or thick-walled capillary scale experiments usually require in the region of 200 μl of solution, and are sometimes most apppropriate as a scale-up from microprocedures.

2.2.3 *Batch*

When conditions for producing crystals are known quite closely, then it is often useful to scale up to a 'batch' method in order to obtain large crystals. This more traditional approach involves adding precipitant or changing pH by careful manual addition of appropriate agents until the protein comes to the limit of its solubility. Following centrifugation or filtration, the sample is stored away in a quiet corner of the laboratory or cold-room for a couple of weeks before inspection. This type of approach was responsible for the production of many of the protein crystals whose study dominated the first decade of protein crystallography, and examples of its application continue to appear. The method is particularly suitable for achieving gradual changes in temperature by insulating the batch container in a water-filled Dewar vessel placed in the cold or warm room. Samples from several ml down to about 100 μl can be employed. It is hoped that by now it will be clear that there is much scope for innovation in the design of crystallizing apparatus and methodology. The novice should try some of the above suggestions with a protein that is cheap and easy to crystallize. For example, lysoszyme will readily crystallize; it should be dissolved at 80 mg ml^{-1} in 40 mM sodium acetate buffer at pH 4.7, slowly mixed with an equal volume of 10% w/v sodium chloride, filtered into a clean glass container and stored for a couple of days at room temperature. It will also crystallize if these conditions are approached gradually by vapour diffusion or dialysis at any scale. Manipulation of the resultant crystals will provide useful practical experience with regard to the mechanical fragility and susceptibility of protein crystals to dehydration. The ease of growing such crystals should not raise too high one's expectations for an early success with a new protein.

The scope for innovation cited above is reflected in the recent description of apparatus for vapour diffusion of drops of protein sandwiched between glass plates, of diffusing precipitants into protein-saturated agarose gels, of the use of microcomputer-controlled motor-driven syringes for reproducible solution handling, and of laboratory robots for setting up vapour-diffusion hanging-drop trays (7).

2.3 **Storage of crystals**

Once crystals have been grown, they can often be stored for extended periods in the mother liquor from which they grew. They can often be transfered to protein-free solution with the same composition as the crystallization cocktail. Crystals grown from high concentration protein solution, but in the absence of precipitants, will often redissolve out of these conditions. They might be stabilized by transfer to solutions containing precipitant which matches the original growth environment, but the choice of conditions must be determined empirically. Extended storage can lead to stabilization of crystals, presumably by photochemical cross-linking reactions. Alternatively, brief (15 min) exposure of crystals to a solution 0.5% in glutaraldehyde can lead to a cross-linked 'skin' over the surface of the crystals which dramatically improves their tolerance to different conditions. It has been observed that crystals that will grow either in ammonium sulphate or polyethylene glycol precipitants are more stable for extended periods in the salt solution, probably because of light-induced degradation of the glycols and the generation of reactive species that damage the protein. Often precipitants will be present at high molarity, and trace contaminants in general laboratory-grade reagents will become significant, so it is wise to use the highest quality available for any reagent which is in contact with the protein. There are no strict limitations on the types of buffer solutions that can be used for crystal growth or storage, but it is worth keeping in mind that compounds like phosphate can produce significant problems of insolubility with heavy metals later in a structure analysis. Furthermore, phosphate crystals readily form from even rather dilute solutions in the presence of agents like methanepentanediol, ethanol or high concentrations of low-molecular-weight glycols, especially at 4°C. Premature celebrations of crystal growth have often occurred when phosphate buffers are employed. As long as one is wary of these pitfalls, phosphate provides a useful wide range of buffer capacity. It has a high solubility in water at room temperature and can be used as a precipitant.

3. FACTORS INFLUENCING CRYSTALLIZATION

3.1 **Effects of impurities**

Lack of purity in the protein sample may lead to (i) complete failure to crystallize, (ii) the production of only small crystals or (iii) large crystals which do not scatter X-rays well. If the protein has not previously been crystallized, then the degree of purity becomes an additional variable to be screened. Contaminating species which are able to satisfy only a portion of the bonding interactions necessary to propagate the lattice may be excluded during crystallization. Any variation in the test molecule in these contact areas will at best reduce the pool of material in the crystallization that can be incorporated in the crystal. In a worst case, where perhaps only a fraction of the crystallization 'valencies' of the protein are damaged, the defective molecules might be built into the lattice and lead to a halt in further growth or disturbances in the lattice. Accumulated defects are believed to be a major factor in limiting crystal size. Crystals of about $0.1 \, mm^3$ are required for diffraction experiments. One implication of this argument is that micro-heterogeneity in the protein of interest is probably more troublesome than contamination from some totally unrelated molecule. This may explain why recrystallization was so useful in some circumstances in purification. The demands of packing in the crystal lattice selected out only those molecules with the appropriate structure. For instance, crystallization of insulin was of major importance in producing higher-quality preparations for the treatment of diabetes mellitus (8). However, we now know that repeated crystallization fails to remove a residual contamination of about 5% comprising proinsulin, and various intermediates of its conversion reaction and chemically modified forms from the extraction process. These components can be removed by gel-filtration, ion-exchange and reverse-phase chromatography. The monocomponent product readily produces large single crystals. Certainly this microheterogeneity presents a more significant purification problem, since variations in properties may be rather slight. It is likely to be encountered to some degree with most protein preparations, reflecting common features of incomplete biosynthetic processing, natural turnover and abuse during extraction.

3.2 **Origins of microheterogeneity**

Microheterogeneity might derive from deamidation of asparagines or glutamines, cysteine oxidation, proteolysis, sequence polymorphism, variability in post-translational modifications like serine phosphorylation or N-methylation or acetylation and sialic acid capping of oligosaccharide chains. Some examples are described.

(i) *Deamidation.* One of the six side-chain amides of bovine insulin deamidates much faster than the others at extremes of pH. The product is fully active and crystallizes isomorphously with undegraded insulin (9). However, the crystal morphology bears little resemblence to the typical rhombohedral form of intact insulin due to a disturbance of the relative growth rates of crystal faces. If an insulin preparation is contaminated heavily with the deamidated form, then curious 'snowflake' polycrystalline aggregates can form in crystallization.

(ii) *Cysteine oxidation.* The reactivity of free thiols varies widely with solvent exposure. Frequently an enzyme active site geometry will enhance the reactivity of such groups which, unless protected, will then be very effective scavengers of trace metals in preparation buffers or readily become oxidized, leading to inactivation and possibly denaturation. The ability of crystals of thymidylate synthase from *Lactobacillus casei* to diffract X-rays has been related to a thiol oxidation state (10). Dimers of reduced or reduced and oxidized forms of the enzyme appear to produce good crystals, while complete oxidation or a random distribution of redox states obtained on exposing preformed crystals to air leads to disorder. Successful control of thiol oxidation was shown to be crucial for reproducible production of crystals of β-hydroxybutyrate dehydrogenase (11).

(iii) *Phosphorylation.* The activity of glycogen phosphorylase is partly controlled by the phosphorylation state of a single serine residue near to the N-terminus. The phosphorylated *a* form is more active. Fortunately, in resting muscle the *a* form predominates, and the enzyme which is responsible for conversion to the *b* form can be quickly removed from extracts. Alternatively, the *b* form may be accumulated. Thus at an early stage the potential difficulties of separating chemically rather similar species are minimized. Glucose-inhibited *a*-form and IMP-activated *b*-form crystals have been studied, and show conformational differences, particularly in the region of the phosphorylation site (12−14).

(iv) *Proteolytic processing.* Preparations of mouse submaxillary gland nerve growth factor show variable chain length following proteolytic processing from a large precursor (15). Eight residues from the N-terminus and one from the C-terminus can be missing. The situation is further complicated since the protein is isolated as a stable dimer with many possible chain-length combinations. The isoelectric points of the variants are very close. The protein has been crystallized, but it remains to be seen whether this heterogeneity has a serious impact on the structure analysis.

(v) *Isoenzymes.* The fungus, *Rhizopus chinensis*, produces two forms of an extracellular aspartic proteinase which have pI values of 5.6 and 6.0. When co-purified, reasonable crystals can be grown. Following separation by preparative isoelectric focusing, the form with pI 5.6 produces larger crystals which diffract X-rays to a higher resolution and are less sensitive to radiation damage (16). The form with pI 6 has not been crystallized.

(vi) *Carbohydrate constituents.* Human leukocyte elastase shows five bands on gel electrophoresis which have been attributed to carbohydrate heterogeneity, but the protein crystallizes well (17). Neuraminidase may be used to remove terminal sialic acid residues from oligosaccharide chains. These residues are probably responsible for the multiple bands seen on gels. Passage of serum transferrin over immobilized neuraminidase followed by separation of the products on chromatofocusing has been reported (18). However, transferrin lacking this treatment has been crystallized (19) and the structure recently determined. Numerous other glycosylated serum proteins have also been successfully crystallized. In spite of this, the carbohydrate attached to proteins is often viewed with suspicion by those trying to grow crystals. This is partly due to its tendency to be heterogeneous, but also perhaps to our poor understanding of the role of carbohydrate. Sometimes the chains are not visible in electron density maps, but in other cases such as immunoglobulins (20), influenza virus coat haemagluttinin (21) and neuraminidase (22), some chains are clearly defined, particularly when involved in protein−protein contact. Both viral proteins are heavily glycosylated (20% by weight). The sugar chains are free of sialic acid, due to the neuraminidase activity. Carbohydrate removal has yet to be shown to be generally useful in crystal growth. Proteins expressed in microorganisms following genetic manipulation will provide a useful source of carbohydrate-free proteins, as complete removal from existing glycoproteins is not easy. Interferon γ provides a recent example (23).

3.3 Detection of microheterogeneity

Microheterogeneity involving a change of a charge in the contact region between molecules in the crystal is likely to be deleterious. Analytical isoelectric focusing on narrow and broad pH ranges together with polyacrylamide gel electrophoresis at pH values above and below the pI are standard procedures capable of detecting most potential problems. Internal chain cleavages may be detected by SDS electrophoresis in reducing conditions. Protein preparations should always be analysed by these methods before crystallization, as high-performance chromatographic supports for ion exchange and chromatofocusing can usually provide the resolution necessary to prepare materials of uniform charge. Silver-staining procedures for analytical electrophoresis methods are in many cases sufficiently sensitive to display purity limits governed by protein stability and handling techniques. A single silver-stained band in a heavily loaded gel is rarely achievable.

Where variability in sequence involves no change of charge, reverse-phase chromatography has proved extremely powerful for smaller proteins. For instance, porcine pituitary gland neurophysin I preparations contain a species in which a single C-terminal leucine deletion occurs, and this was first detected by reverse-phase chromatography (24). Similarly, bovine pancreatic proinsulin exhibits a single Leu−Pro polymorphism which is resolvable (25). Of course, at the beginning of crystallization trials one cannot predict how damaging a particular contaminant might be. The various types of heterogeneity outlined above may fortuitously fall outside contact regions between molecules and be tolerated in the lattice. Neurophysin and proinsulin both crystallize fairly well (26,27) without the separation of the sequence variants, while the two insulins of the rat do not, since they crystallize in the same conditions to produce a rhombohedral and a cubic packing arrangement (28).

Heterogeneity may, on the other hand, accompany partial unfolding of the protein or promote distinct conformational isomers in more flexible proteins which will not be so readily detected. Denaturation might only be observed as a change of solubility, a slight alteration in apparent molecular size during gel-filtration chromatography or an increased susceptibility to proteolysis. Freeze-drying was found to be detrimental to the crystallization of carbonic anhydrase, presumably through partial denaturation (29). Many enzymes need to be protected from the denaturing effects of ice crystals by storage at low temperatures in glycerol.

Where a difficult final purification step is to be considered, with the risk of a further decline in yield of a scarce material, one must try to equate the benefits with the cost. It is probably not unreasonable to suppose that the ability of the crystal lattice to select out suitable molecules is a common feature of protein crystals, the degree of success reflecting the variable strength of lattice contacts. For this reason, it is worthwhile attempting to grow crystals from a number of preparations whose purity approaches the ideal.

Preparations of pituitary growth hormone of variable quality have been available for almost 40 years, and many attempts at crystallization have failed. Fresh human tissue was of course rarely available, and traditional extraction methods from glands of various animals involved a number of rather harsh steps in order to cope with the troublesome solubility properties of growth hormones. Most preparations from glands show at least two bands on isoelectric focusing, and this was attributed to partial removal of the amino-terminal residue. The pituitary gland produces, in addition to 22-kd growth hormone, a form in which 2 kd of sequence is omitted from within the polypeptide chain. Covalent dimerization has also been observed. Differential deamidation rates have been observed for these species. Recently, bovine growth hormone produced by milder purification methods has produced small crystals (30), and good crystals of porcine and human growth hormone synthesized in *E.coli* by recombinant DNA methods have been produced (31,32). The tendency of human growth hormone to aggregate through non-specific hydrophobic interactions was successfully controlled using the detergent *n*-octyl-β-D-glucopyranoside. Otherwise, the crystals are grown under rather straightforward conditions, indicating that improvement in product quality has been the decisive factor in initiating the structure analysis.

On occasions, the subtleties of a particular state of purity may not be evident (for example in the exact lipid−detergent composition associated with a solubilized membrane protein complex), and in such cases rigorous reproduction of purification conditions is essential, particularly once crystals have been grown. In fact, rather few membrane proteins have been successfully crystallized. The methods used are very similar to those outlined in Section 2, but detergent is an obligatory constituent. Other small amphiphilic additives are also used (33−36).

One should also be wary of the colour of a protein preparation. If it is coloured without good reason, then it is either impure or one is on the brink of some discovery. With large proteins, it may not be possible to detect some of the variables easily seen in smaller ones, but hopefully the potential for damage in crystallization is diluted to a similar degree.

3.4 Additives and proteolysis

In contrast to the previous sections, there are many examples in which non-peptide

55

species other than precipitants are important in crystallization. Cofactors, allosteric effectors and metal ions required for enzyme activity are perhaps the easiest to identify as specific activity is often monitored to follow purification. Furthermore, such compounds can be protective and are usefully included in preparation buffers. Enzyme substrates, products and carrier protein ligands have also been useful. Malic enzyme from rat liver, for instance, requires NADP for crystallization (37). Enzyme inhibitors have been employed in the crystallization of *E.coli* dihyrofolate reductase (38) and mouse submaxillary gland renin (39). The rate of crystal growth of serum amyloid protein can be controlled by the inclusion of a modified monosaccharide similar to that employed as an immobilized ligand on affinity-chromatography supports during purification (40). Some caution is necessary, as binding of allosteric effectors to phosphofructokinase and aspartate carbamoyl transferase leads to substantial conformation change and subunit rearrangements (41,42). Many enzyme crystals disintegrate in the presence of substrates and modifiers, indicating that the protein conformation is no longer consistent with the existing set of intermolecular contacts. Substantial changes in crystal form are found for various apo- and ternary complex forms of lactate dehydrogenase (43). Glyceraldehyde-3-phosphate dehydrogenase can be prepared with from 0 to 4 molecules of NAD bound (44). The crystallization of concanavalin A is inhibited by its ligand *N*-acetyl glucosamine. It is hard to find guiding general principles within this array of experiences. It is clear that partial occupation of a protein preparation with such compounds could be a serious source of microheterogeneity. However, once under control, their binding provides the opportunity to investigate the relation of structure and mechanism.

In other cirucmstances, purification may unwittingly remove some essential component for crystallization. Chelating agents or dissociating conditions are often used to aid in combating proteolysis in crude tissue extracts, and these will remove metals required for crystal growth. The requirement for zinc ions in the crystallization of insulin is a well-known example. Copper ions were found eventually to be essential for the crystallization of oxidized thioredoxin, although they were not known to be involved with the activity of the molecule (45). Cobalt ions were necessary to crystallize chloramphenicol acetyl transferase (46). In these examples, the metal ions seem to be important only in providing a link between molecules in the crystal lattice.

There are many examples (1) where controlled proteolysis leads to a species that is particularly suitable for crystallization (for example immunoglobulins, canavalin, elongation factor Tu, ribonuclease S, cytochrome b5). In some cases, it seems that proteolysis is removing flexible or protruding portions of the protein to provide more compact structures which are easier to pack together. In other cases, linking strands of polypeptide can be cleaved to release intact domains. The ionophore domain of colicin A (47) and the Klenow fragment of DNA polymerase (48) are produced by thermolysin and subtilisin digestion for crystallization. Influenza virus haemagglutinin and neuraminidase provide interesting examples. Both native proteins are attached to the viral envelope membrane by a hydrophobic domain which is cleaved during preparation by bromelain or pronase! Their heavy glycosylation probably protects against more extensive proteolysis.

4. CONCLUDING REMARKS

So far we have noted how microheterogeneity can have deleterious effects for the crystal lattice and it is not hard to envisage how the solubility properties of the protein might be blurred by such variability, leading to difficulties in defining crystallization conditions. For instance, traces of a contaminant or modified form of the protein of interest might be precipitated during a vapour diffusion equilibration before crystallization starts, producing an excess of nucleation sites. If crystals are obtained in spite of these factors, further problems may be waiting. Crystal morphology may be variable due to minor packing changes and resultant changes in the rate of development of crystal faces. The efficiency of heavy-metal binding to the protein which is necessary for structure analysis might be impaired if, for instance, proteolysis or chemical modification has effectively removed important residues from a significant proportion of the protein molecules. Such degradation might also lead to poor definition in this part of the final structure. Correlations of structure and function may be led astray if chemical differences from the native form are not fully appreciated.

Crystal growth is not well understood, and almost entirely empirical rules are followed to grow crystals. Criteria of purity for crystallization should in practice be no more or less stringent than those tolerated for any other means of characterization of a pure protein. Inasmuch as some methods of study need not necessarily be harmed by some lack of purity, the same may often be true of crystallization. However, the basis of the technique of X-ray analysis presumes identity of the molecules under study, and towards the end of a structure analysis the extent of any compromise at an early stage may be magnified to cause serious difficulties in interpretation.

Wherever rather special and complex recipes are required for crystal growth it is unusually hard to outline a logical path to the eventual conditions or to see how systematic screening could cope alone. Rather, screening with a strong bias from prior knowledge of the properties of the protein gained during isolation seems the most powerful approach.

5. REFERENCES

1. McPherson,A. (1982) *Preparation and Analysis of Protein Crystals*. John Wiley, New York.
2. Blundell,T.L. and Johnson,L. (1976) *Protein Crystallography*. Academic Press, New York.
3. Arakawa,T., Timasheff,S.N., Feher,G., Kam,Z. and McPherson,A. (1985) In *Methods in Enzymology*. Wyckoff,H.W. *et al.* (eds), Academic Press, New York, Vol. 114, p. 19.
4. Gilliland,G.L. and Davies,D.R. (1984) In *Methods in Enzymology*. Colowick,.S.P. *et al.* (eds), Academic Press, New York, Vol. 104, p. 370.
5. Zeppenzauer,M., Eklund.H. and Zeppenzauer,E. (1968) *Arch. Biochem. Biophys.*, **126**, 564.
6. Wlodawer,A., Hodgson,K. and Shooter,E. (1975) *Proc. Natl. Acad. Sci. USA*, **72**, 777.
7. Proceedings of the Second International Conference on Protein Crystal Growth, Strasbourg (1987) *J. Cryst. Growth*, **90**, 1.
8. Blundell,T.L., Dodson,G.G., Hodgkin,D.C. and Mercola,D.A. (1972) *Adv. Protein Chem.*, **26**, 279.
9. Bedarkar,S. (1982) PhD. Thesis, University of London.
10. Tykarska,E., Lebioda,L., Bradshaw,T.P. and Dunlap,R.B. (1986) *J. Mol. Biol.*, **191**, 147.
11. Drenth,J. (1988) *J. Cryst. Growth*, **90**, 368.
12. Cori,G.T., Illingworth,B. and Keller,P.J. (1955) In *Methods in Enzymology*. Colowick,S.P. and Kaplan,N. (eds), Academic Press, New York, Vol. 1, p. 200.
13. Fletterick,R.J., Sprang,S. and Madsen,N.B. (1979) *Can. J. Biochem.*, **57**, 789.
14. Weber,I.T., Johnson,L.N., Wilson,K.S., Yeates,D.G.R., Wild,D.L. and Jenkins,JA. (1978) *Nature*, **274**, 433.
15. Server,A.C. and Shooter,E.M. (1977) *Adv. Protein Chem.*, **31**, 339.
16. Bott,R.R., Navia,M.A. and Smith,J.L. (1982) *J. Biol. Chem.*, **257**, 9883.
17. Bode,W., Wei,A.-Z., Huber,R., Meyer,E., Travis,J. and Neumann,S. (1986) *EMBO J.*, **5**, 2453.

18. *FPLC Ion Exchange and Chromatofocusing—Principles and Methods*. Laboratory Separations Division, Pharmacia AB, Uppsala, 1985.
19. Al-Hilal,D., Baker,E., Carlisle,H., Gorinsky,B., Horsburgh,R.C., Lindley,P.F., Moss,D., Schneider,H. and Stimpson,R. (1976) *J. Mol. Biol.*, **108**, 255.
20. Deisenhofer,J., Colman,P.M., Epp,O. and Huber,R. (1976) *Hope-Seyler's Z. Physiol. Chem.*, **357**, 1421.
21. Wilson,I.A., Skehel,J.J. and Wiley,D.C. (1981) *Nature*, **289**, 366.
22. Varghese,J.N., Laver,W.G. and Colman,P.M. (1983) *Nature*, **303**, 35.
23. Vijay-Kumar,S., Senadhi,S.E., Ealick,S.E., Nagabhushan,T.L., Trotta,P., Kosecki,R., Reichert,P. and Bugg,C.E. (1987) *J. Biol. Chem.*, **262**, 4804.
24. Schwandt,P. and Richter,W.O. (1980) *Biochim. Biophys. Acta*, **626**, 376.
25. Frank,B.H., Pekar,A.H., Pettee,J.M., Schirmer,E.M., Johnson,M.G. and Chance,R. (1984) *Int. J. Peptide Protein Res.*, **23**, 506.
26. Pitts,J.E., Wood,S.P., Hearn,L., Tickle,I.J., Wu,C.W., Blundell,T.L. and Robinson,I.C.A.F. (1980) *FEBS Lett.*, **121**, 1.
27. Blundell,T.L., Pitts,J.E. and Wood,S.P. (1982) *Crit. Rev. Biochem.*, **13**, 141.
28. Wood,S.P., Tickle,I.J., Bludell,T.L., Wollmer,A. and Steiner,D.F. (1978) *Arch. Biochem. Biophys.*, **186**, 175.
29. Kannan,K.K., Fridborg,K., Bergsten,P.C., Liljas,A., Lovgren,S., Petef,M., Strandberg,B., Waara,I., Adler,L., Falkenbring,S.O., Gothe,P.O. and Nyman,P.O. (1972) *J. Mol. Biol.*, **63**, 601.
30. Bell,J.A., Moffat,K., Vonderhaar,B.K. and Golde,D. (1985) *J. Biol. Chem.*, **260**, 8520.
31. Abdel-Meguid,S.S., Smith,W.W., Violand,B.N. and Bentle,L.A. (1986) *J. Mol. Biol.*, **192**, 159.
32. Jones,N.D., DeHoniesto,J., Tackitt,P.M. and Becker,G.W. (1987) *Bio/Technology*, **5**, 499.
33. Michel,H. (1982) *J. Mol. Biol.*, **158**, 567.
34. Garavito,R.M., Jenkins,J., Jansonius,J.N., Karlsson,R. and Rosenbusch,J.P. (1983) *J. Mol. Biol.*, **164**, 313.
35. Michel,H. and Oesterhelt,D. (1980) *Proc. Natl. Acad. Sci. USA*, **77**, 1283.
36. Gros,P., Groendjik,H., Drenth,J. and Hol,W.G.J. (1988) *J. Cryst. Growth*, **90**, 193.
37. Baker,P.J., Thomas,D.H., Howard Barton,C., Rice,D.W. and Bailey,E. (1987) *J. Mol. Biol.*, **193**, 233.
38. Mathews,D.A., Alden,R.A., Bolin,J.T., Freer,S.T., Hamlin,R., Xuong,N., Kraut,J., Poe,M., Williamson,M. and Hoogsteen,K. (1977) *Science*, **197**, 452.
39. Navia,M.A., Springer,J.P., Poe,M., Bojer,J. and Hoogsteen,K.J. (1984) *J. Biol. Chem.*, **259**, 12714.
40. O'Hara,B.P., Wood,S.P., Oliva,G., White,H.E. and Pepys,M.B. (1988) *J. Cryst. Growth*, **90**, 209.
41. Evans,P.R., Farrantes,G.W. and Lawrence,M.C. (1986) *J. Biol. Chem.*, **191**, 713.
42. Krause,K.L., Volz,K.W. and Lipscomb,W.N. (1985) *Proc. Natl. Acad. Sci. USA*, **82**, 1643.
43. Schar,H.-P., Zuber,H. and Rossman,M.G. (1982) *J. Mol. Biol.*, **154**, 349.
44. Skarzynski,T., Moody,P.C.E. and Wonacott,A.J. (1987) *J. Mol. Biol.*, **193**, 171.
45. Soderberg,B.-O., Holmgren,A. and Branden,C.-I. (1974) *J. Mol. Biol.*, **90**, 143.
46. Leslie,A.G.W., Moody,P.C.E. and Shaw,W.V. (1988) *Proc. Natl. Acad. Sci. USA*, **85**, 4133.
47. Tucker,A.D., Pattus,F. and Tsernoglou,D. (1986) *J. Mol. Biol.*, **190**, 133.
48. Ollis,D.L., Brick,P., Hamlin,R., Xuong,N.G. and Steitz,T.A. (1985) *Nature*, **313**, 762.

CHAPTER 4

Purification of membrane proteins

J.B.C.FINDLAY

1. INTRODUCTION

The single most important feature to ascertain about a membrane protein at the outset is its mode of association with the bilayer (*Figure 1*). This information will largely govern the methods to be adopted for its purification. If integrated into the hydrophobic phase, liberation of the so-called *integral* (or *intrinsic*) membrane proteins in soluble form will require disruption of the phospholipid bilayer or cleavage of the polypeptide from its membrane anchor. During all the subsequent purification procedures, 'shielding' of any substantial intramembranous hydrophobic face from the aqueous media will be required; this is usually achieved with detergents. *Peripheral* (or *extrinsic*) membrane proteins, on the other hand, are associated with the membrane surface through interactions either with other proteins or with the exposed regions of phospholipid. In these cases, the protein may often be dissociated from the membrane by altering the ionic conditions of the buffer or, in more resilient situations, by inducing a degree of denaturation. Although detergents can help liberate this type of protein from the

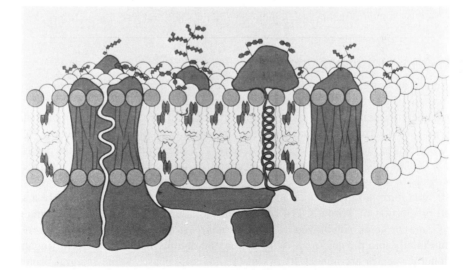

Figure 1. Schematic diagram of generalized membrane structure. The membrane proteins can be categorized into two classes: integral (see *Figure 2*) and peripheral. The latter group are considered not to make hydrophobic contacts with the bilayer. Both groups of proteins can exist as monomers or as hetero- or homo-oligomers. Carbohydrate residues (♦) are shown on the extracellular surface of the membrane.

membrane, they are rarely required subsequently.

Since membrane proteins normally represent a very small proportion of total cellular protein, any contamination is going to have a disproportionate effect. For this reason, the most valuable initial step is to obtain as pure a preparation of the appropriate membrane as possible. A complete guide to membrane isolation is given in Chapters 1 and 2 of ref. 1. Starting with the whole cell or tissue, the first important point to establish is the best procedure for generating the membrane under study. Some cells such as erythrocytes and tumour cells can be persuaded to shed, without generalized lysis, microvesicles representative of the plasma membrane or containing a discrete subset of its proteins. Other, usually highly differentiated, systems can be treated so that a specific section of the cell can be split off. A good example of this is the visual retina of both vertebrates and invertebrates, where the membrane (multi-invaginated or microvillar) containing the primary photoreactive apparatus can be liberated almost intact from the cell by very gentle agitation. Such procedures enormously simplify both the number and complexity of the steps that are otherwise normally required when dealing with tissue subjected to aggressively disruptive techniques.

Separation of membrane fractions or subcellular organelles can be achieved by a number of different approaches. The most successful of these is centrifugation, normally using a density gradient of some kind. Other approaches which have been used to advantage include two-polymer phase partitioning, high-voltage free-flow electrophoresis, and (immuno)affinity and gel-filtration chromatography. Considerable patient application and ingenuity has gone into membrane isolation, and good methods now exist for most organelles and membrane systems. At this early stage in our study of membrane proteins, therefore, much is to be gained by careful attention to the extensive literature on membrane purification. The remainder of this chapter will be concerned with describing either methods devised specifically for dealing with membrane proteins, or standard techniques which have been modified to cope with this class of polypeptide.

2. PROTEIN SOLUBILIZATION

There are a number of ways in which one can relatively quickly determine whether the protein of interest is a member of the integral or peripheral fraternity. Most involve subjecting the membrane to the washing or stripping procedures identified below. Charge-shift electrophoresis has also been successfully employed (see Section 3.2.1), but a newer and more rapid procedure makes use of phase partitioning in the detergent Triton X-114 (1,2). The particular advantage of this detergent is that its micelles aggregate on warming from $0°C$, eventually separating out into a second phase when the temperature is raised above $20-25°C$ (the so-called 'cloud point'). Only a very small proportion of Triton X-114 remains in the aqueous phase. Therefore, integral membrane proteins solubilized at $0-4°C$ tend, at the 'cloud point', to partition preferentially into the detergent-rich fraction and can be recovered by centrifugation. Peripheral species are almost always in the aqueous phase. Some integral proteins, particularly those with pronounced hydrophilic character (for example proteins with only a single intramembranous portion and/or with considerable attached carbohydrate) can sometimes remain in the detergent-poor phase. This also appears to be the case for some transport proteins, for example the acetylcholine-activated ion-channel and

Method Table 1. Phase-partitioning with Triton X-114

A. Detergent purification

1. Add 10 g of Triton X-114 (Sigma or Calbiochem) and 8 mg of butylated hydroxytoluene to a final volume of 200 ml in 20 mM potassium phosphate pH 7.5, 0.15 M KCl.
2. Cool the mixture to near 0°C, remove any insoluble material by centrifugation (3000 *g* for 10 min), then incubate the soluble detergent for 18−20 h at 35°C.
3. Centrifuge gently (100 *g* for 5−10 min) and retain the detergent (lower) phase. Alternatively, the detergent will settle out without centrifugation.
4. Steps 1−3 can be repeated if necessary. The purified detergent may be stored at room temperature for at least 3 months.

B. Membrane solubilization and partitioning

1. Add purified Triton X-114 to membranes at 0−4°C in 10 mM Tris−HCl or phosphate buffer, pH 7.4 containing up to 150 mM NaCl or KCl. The final detergent concentration can be up to 2−3% w/v and the membrane protein 1−3 mg ml^{-1}.
2. Incubate at 0−4°C for up to 1 h then centrifuge the mixture (100 000 *g*, 1 h, 0−4°C) to remove insoluble material. Check that the protein of interest is not insoluble at this stage.
3. Overlay the soluble fraction on a cushion of buffered 6% sucrose in a suitable centrifuge tube.
4. Warm to 25−30°C for 5−10 min, then centrifuge for 5−10 min at 100 *g* and above 20°C.
5. The detergent-rich phase appears as an oily layer at the bottom of the tube. The aqueous fraction can be re-equilibrated with 0.5% (w/v) Triton X-114 and the process repeated. Similarly the detergent phase can be re-extracted with aqueous buffer.

certain ATPases (3,4), although the reasons for this behaviour are unclear. Procedures for purifying Triton X-114 and for carrying out a phase-partitioning experiment are given in *Method Table 1*. Before embarking on this approach, it may be important to ensure that the protein of interest retains either its biological activity, or at least some other means of identification, when incubated with the detergent.

2.1 Peripheral membrane proteins

There are a significant number of treatments to which the membrane can be subjected in order to remove the peripheral components. Mild conditions include low-ionic-strength buffers, 0.1−1 mM EDTA (5,6) or high-ionic-strength solutions with or without up to 1 M NaCl or KCl (6,7). To prevent irreversible protein denaturation, the pH should be in the range 6−8 and incubations should be carried out in the cold. Anions, such as iodide and diiodosalicylate, are to be avoided since, being chaotropic agents, they have detergent-like solubilizing properties. In all manipulations of the membrane, the inclusion of protease inhibitors [see the companion volume (8) and ref. 1] should be considered if only because some of the treatments, for example EDTA and thiol reagents, may activate proteolytic enzymes. A useful, but not comprehensive, cocktail of inhibitors includes 100 μM phenylmethylsulphonyl fluoride (PMSF), 1 mM sodium tetrathionate and 1 mM 2-phenanthroline. PMSF is toxic and should therefore be handled with care. It should be remembered that this reagent has a half-life in aqueous solutions at 25°C of only 35 or 110 min at pH 8.0 and 7.0 respectively. It is usually stored at 100 mM in 2-propanol or ethanol at -20°C.

Many peripheral membrane proteins are not liberated by these milder conditions but require harsher treatments which usually result in protein denaturation, often irreversible. Such aggressive conditions include 6 M guanidine hydrochloride (9), 8 M urea, 1 mM *p*-chloromercuribenzoate (10), dilute acids, pH 2.0−3.0 (11) and dilute alkali, pH 9.5−11.0 (12). The alkali treatment is preferred to acidic conditions, since the latter sometimes results in precipitation of protein which interferes with subsequent membrane recovery. To prevent cleavage of peptide bonds, 'stripping' of the membranes at extreme pH is usually carried out at 0−4°C, and neutrality restored as soon as possible.

As a result of these various treatments, 50% or more of the total protein may be released from the membrane. Very often this causes vesiculation, fragmentation or inversion of the membrane. The recovery conditions, centrifugation, filtration, etc., should, therefore, be altered to ensure maximum yield of membrane.

2.2 Integral membrane proteins

Integral membrane proteins may themselves be classified into four groups depending on the proportion of their structure that is in contact with the hydrophobic phase of the bilayers (see *Figure 2*). One group is anchored into the membrane by one or two

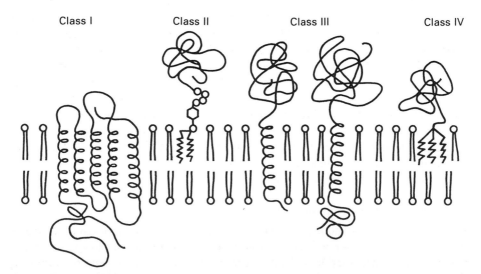

Figure 2. Structural types of integral membrane proteins. Class I, proteins in which a substantial proportion of their mass is embedded in the bilayer; Class II, proteins bound to the bilayer via a phosphatidyl inositol-carbohydrate moiety attached to the C terminus of the polypeptide chain; Class III, proteins anchored in the bilayer by a single transmembrane segment; and Class IV, proteins associated with the bilayers via fatty acyl and/or diacylglycerol moieties covalently attached to the N terminus of the polypeptide chain. The latter three categories of protein contain most or all of their mass in the aqueous phase.

transmembrane protein segments (13) while a second class utilizes covalent attachment to fatty acids or lipid (14,15). The bulk of the protein in both these groups is in the aqueous phase and in general their biological activity does not involve the lipid bilayer. Examples include a large number of enzymes such as peptidases, esterases and phosphatases. In these instances, the bilayer simply acts as a structural support or means of localization and/or organization. Proteins of this type can often be liberated from the membrane by the use of proteases or phospholipases without destroying their native activities. The role of the cleavage enzymes is to separate the protein from its covalently attached protein or lipid membrane anchor which remains behind in the bilayer.

The phosphatidyl inositol phospholipase C from *Bacillus cereus, B.thuringiensis* or *Staphylococcus aureus* is thought to cleave the attached phospholipid such that a diacylglycerol moiety is left in the bilayer, while inositol phosphate together with an intervening glycan segment remains covalently associated with the solubilized protein. Recently it has been reported that a phospholipase D present in serum is capable of liberating inositol-linked alkaline phosphatase from the membrane, leaving behind the phosphatidic acid fraction (16). Typical protocols are given in *Method Table 2*. The efficiency of any particular enzyme is dependent on the membrane protein under investigation. The incubation conditions have to be carefully chosen so as not to produce proteolysis within the globular portion of the membrane protein.

A third, and so far relatively small but nevertheless very important, group of proteins also possesses only a single membrane-spanning segment. Receptors such as those for insulin, growth factors, and LDL (19,20), are representative of this class. They differ from the first group in having substantial portions of their mass on both membrane surfaces. Their full biological activity is dependent not only on the two hydrophilic domains, but also on the transmembrane segment. However, the globular portions in the aqueous phase can be freed from the rest of the molecule by proteolytic action as described above. Under these conditions it is still possible for the extracellular domain, for example, to retain its ligand-binding activity. Full biological activity is lost, however, for there is now no means of communicating with the intracellular domain.

Members of the fourth major class of integral membrane proteins traverse the bilayer several times and have a major fraction of their total structures within the hydrophobic phase. Transport proteins and receptors of the rhodopsin type fall into this group (21). Although proteolysis can release fragments of such proteins into the aqueous phase (thereby often destroying their biological activities), solubilization from the bilayer almost always requires organic solvents or detergents.

2.2.1 *Delipidation by organic solvents*

Although most types of phospholipid are readily soluble in the majority of pure organic solvents, the same is not necessarily true for most integral membrane proteins. The reasons for this appear to lie in the amphipathic nature of their structures. Protocols which have been used involve treatment of lyophilized membrane (100−200 mg) with 100 to 200 ml of solvent: (i) chloroform alone or with methanol, ethanol or butanol in various proportions (3:1 to 1:3 v/v); (ii) butanol on its own or in combination; or (iii) mixtures of diethyl ether and ethanol. Under these conditions, most membrane proteins promptly precipitate (22−25). Only a small proportion is soluble in the organic

Method Table 2. Incubation conditions for proteases and phospholipase.

A. Proteases

A wide range of proteases is available for this type of study. Amongst the more successful have been plant enzymes such as papain, ficin and bromelain, the enzyme from *Staphylococcus aureus* V8 (Glu-specific), and endoprotease Lys-C. Depending on the particular protein and membrane, however, other proteases, such as trypsin, chymotrypsin and subtilisin, may be more useful.

1. Suspend membrane (1 mg protein ml^{-1}) in suitable buffer (0.1 M phosphate, pH 7.0 for V8 protease; 0.1 M phosphate or Tris buffer, pH 7.0−8.6 for other proteases).
2. Add up to 2% w/w of the proteolytic enzyme.
3. Incubate at 25−37°C for up to 3 h.
    Either
4. Take aliquots at various time intervals into the SDS-containing loading buffer for polyacrylamide gels (see ref. 17). Protease inhibitors can be added to ensure termination of digestion.
    Or
5. Terminate reaction with inhibitors, spin down membrane and examine supernatant for presence of the protein.

B. Phospholipases

A very similar approach can be used to examine whether the protein is linked via a phosphatidyl inositol moiety. The enzyme, PI-phospholipase C, should be as pure as possible since contaminating proteases, etc., can yield artefactual results. For these reasons, it is often best to purify the enzyme (18). However, success can also be obtained with the phospholipase C preparation from Sigma which possesses PI-phospholipase C activity in addition to other less specific phospholipases. Typical reaction conditions are:

> 0.1 M Tris−HCl, pH 7.4
> 25−30°C for up to 3 h
> 1−2% w/w enzyme or 0.1−1 μg ml^{-1}

NaCl produces inhibition and should be avoided. EDTA can be present.

---

medium. The spectrum of protein extracted can vary slightly with solvent composition (26). However, they are all characterized by pronounced hydrophobic character and have been designated proteolipid proteins, for example those from nerve myelin (22) and the dicyclohexylcarbodiimide (DCCD)-binding protein from the ATP synthase complex (27), both of which are soluble in chloroform:methanol 2:1 (v/v).

More often the need is to extract aqueous membrane preparations, when the following procedure can be used.

(i)   Mix membranes (1−2 mg protein ml^{-1}) in buffer with 4 to 5 times the volume of cold chloroform:methanol (2:1 v/v).

(ii)   Shake vigorously and centrifuge (3000 *g* for 10−20 min).

(iii) Three layers are usually observed: a lower organic layer which contains most of the phospholipids, cholesterol and proteolipid proteins; a substantial intermediate (interfacial) layer of precipitated protein; and an upper aqueous layer which contains particularly hydrophilic species such as heavily glycosylated proteins (e.g. glycophorin and the larger and more complex glycolipids). The addition of salt (e.g. 0.1 M KCl) helps to ensure that most of the lipids partition into the organic phase.

The use of acidified organic solvents, such as 50 mM HCl or 5 mM *p*-toluene sulphonic acid in chloroform:methanol (2:1 v/v), can markedly increase the range and amount of soluble protein obtained (28,29), but at the cost of almost inevitable protein denaturation. Chloral hydrate is also a successful solvent, but at the high concentrations used (up to 80%), its toxicity becomes a serious hazard. Most success has been obtained with formic acid−acetic acid−chloroform−ethanol or FACE (1:1:2:1 v/v; 30,31). Added to pelleted membranes, where the volume of water is small and readily accommodated by the solvent, complete or almost complete solubilization of the membranes can be achieved. Under these conditions, both the lipid and protein components are released in monomeric, un-aggregated form. Subsequent fractionation in this solvent using appropriate resins can then be carried out (see Section 3.1).

There are a number of ways in which solubilized proteins can be recovered. Removal of solvent under a stream of nitrogen (in a fume-hood) or by rotary evaporation are the simplest, but in both cases the lipids are still present and some proteins can become difficult to handle (for instance, aggregate irreversibly). Alternatively, it is possible to precipitate them out using 10−20 vols of cold acetone, 5−10 vols of cold diethyl ether or several vols of diethylether−ethanol mixtures (1:1 to 3:1 v/v). Perhaps the best way, however, especially if further purification is envisaged, is to employ some form of column chromatography. Gel filtration using Sephadexes LH60 and LH20 (25,32), ion-exchange chromatography with CM− and DEAE−cellulose (33,34) and reverse-phase chromatography using silica (35) have all been successful.

Once free of lipid, the proteins, or hydrophobic peptides therefrom, can be dissolved in formic acid mixtures (32,36) with water (30−90%), ethanol (3:7 or 1:2 v/v) or isopropanol (up to 1:2 v/v). Polyacrylamide resins such as the Biogel P series can be used with the first of these solvents, and Sephadexes LH60 and LH20 and the reverse-phase silicas with the latter two. These formic acid solutions have some advantages with protein−peptide preparations from which most of the lipid and detergent has been removed.

Proteins or peptides can be recovered from these acidic-organic conditions by gentle rotary evaporation to a wet film, followed by freeze-drying from aqueous detergent solutions. Complete removal of acidic-organic solvents often leads to irreversible aggregation if the proteins are then exposed to neutral aqueous conditions. In such situations, it is best to add the detergent before the sample is completely dried.

2.2.2 *Chaotropic agents*

Chaotropic agents increase the solubility of non-polar moieties in aqueous solvents. Their mechanism of action is related to their ability to 'structure' water. Because of this property they are more useful in selectively extracting proteins from the membrane rather than in dissolving the entire bilayer. Yields are generally poor. Their efficacy declines in the order perchlorate, thiocyanate, guanidine, chlorate, iodide, bromide, nitrate and urea. Lithium diiodosalicylate is probably about the most successful member of this group and has been used in the purification of highly glycosylated integral membrane proteins (37).

2.2.3 *Use of detergents*

Detergents are amphiphilic, usually relatively small, molecules which are soluble in aqueous conditions on account of their hydrophilic moieties (Chapter 3, ref. 1; 38−41). Above a certain concentration (called the critical micellar concentration, or CMC), the molecules coalesce via their hydrophobic portions into thermodynamically stable aggregates (called micelles) which are in equilibrium with the free monomer. When added to membranes they rapidly partition into the bilayer until a point is reached whereby the membrane is no longer stable and begins to disintegrate. If sufficient amounts of a suitable detergent have been added, the membrane is eventually 'solubilized' into its lipid and protein components. The former generally form mixed micelles with the amphiphile, while the latter exist in solution surrounded by a monomolecular layer of detergent which intercedes between the hydrophobic surface of the protein and the bulk aqueous medium. This represents a generalized scheme of events since different membranes and different proteins respond in different ways to different detergents. A number of factors, therefore, have to be considered in the design of a solubilization medium.

(i) *Detergent properties and selection.* The relevant structural characteristics and behaviour of a number of readily available detergents are given in *Figure 3* and *Table 1*. It is quickly apparent that they vary widely, but a less obvious feature is that any particular preparation may be heterogeneous or polydisperse, for example in the hydrocarbon chain length. The purer the detergent, therefore, the more reproducible its properties will be. In general, it is best to use as pure a detergent as possible, and purities close to 100% can be obtained. However, since quality varies from supplier to supplier and from batch to batch, it is difficult to give definitive advice other than to suggest that increased purity often goes hand in hand with increased price. It is also recommended that detergents be purchased in the minimum quantities required, since deterioration can result from prolonged storage. The CMC for non-ionic detergents decreases with increasing temperature (42), as do those of ionic detergents with increasing salt concentrations (43). It is not usually possible to predict how any particular protein will respond to any particular detergent. However, one can produce a number of general guidelines which might focus the choice.

(a)   The most effective detergents tend also to be the most aggressive and the least likely to sustain biological activity. At one extreme are SDS and CTAB, which are good solubilizing agents but usually also denature the protein. However, a recent interesting development is the synthesis of sodium oligooxyethylene dodecyl ether sulphates which have good solubilizing properties, yet preserve protein activity, and can be used successfully in polyacrylamide gel electrophoresis

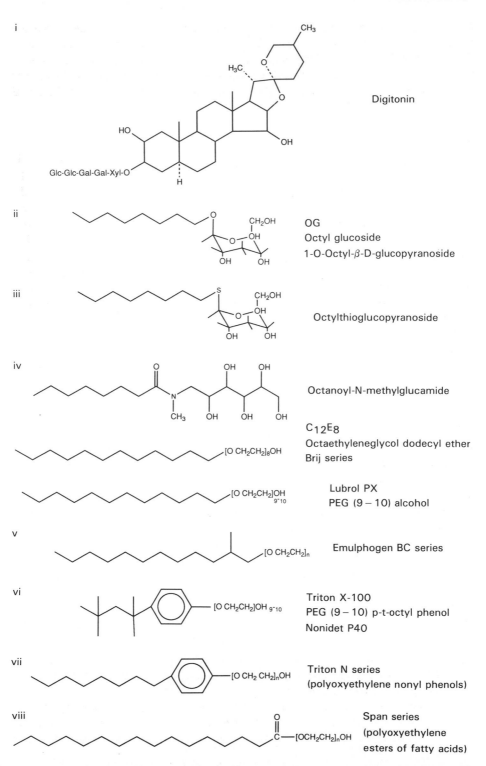

i

CH_3

H_3C

Digitonin

HO

OH

Glc-Glc-Gal-Gal-Xyl-O

H

ii

O

CH_2OH

OH

O

OH

OH

OG
Octyl glucoside
1-O-Octyl-β-D-glucopyranoside

iii

S

CH_2OH

OH

O

OH

OH

Octylthioglucopyranoside

iv

O

OH

OH

N

CH_3

OH

OH

OH

Octanoyl-N-methylglucamide

$[O\ CH_2CH_2]_8OH$

$C_{12}E_8$
Octaethyleneglycol dodecyl ether
Brij series

$[O\ CH_2CH_2]OH_{9-10}$

Lubrol PX
PEG (9 – 10) alcohol

v

$[O\ CH_2CH_2]_n$

Emulphogen BC series

vi

$[O\ CH_2CH_2]OH_{9-10}$

Triton X-100
PEG (9 – 10) p-t-octyl phenol
Nonidet P40

vii

$[O\ CH_2\ CH_2]_nOH$

Triton N series
(polyoxyethylene nonyl phenols)

viii

O

C

$[OCH_2CH_2]_nOH$

Span series
(polyoxyethylene
esters of fatty acids)

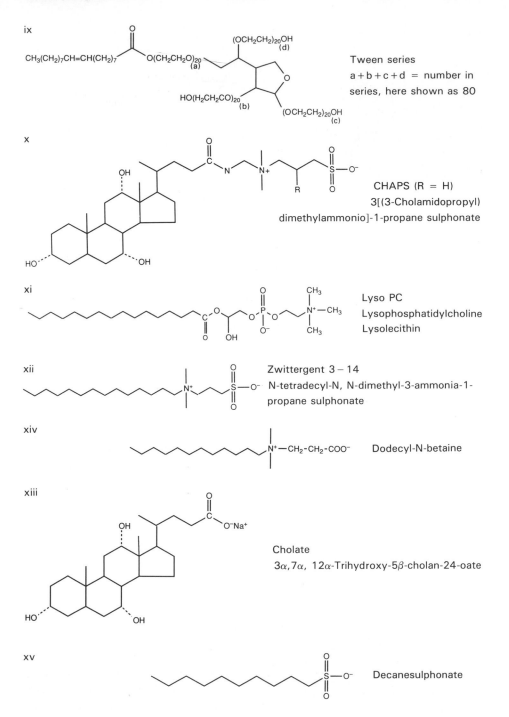

ix

$(OCH_2CH_2)_{20}OH$
(d)

$CH_3(CH_2)_7CH=CH(CH_2)_7$ $O(CH_2CH_2O)_{20}$
(a)

$HO(H_2CH_2CO)_{20}$
(b)

$(OCH_2CH_2)_{20}OH$
(c)

Tween series
a + b + c + d = number in
series, here shown as 80

x

CHAPS (R = H)
3[(3-Cholamidopropyl)
dimethylammonio]-1-propane sulphonate

xi

Lyso PC
Lysophosphatidylcholine
Lysolecithin

xii

Zwittergent 3 – 14
N-tetradecyl-N, N-dimethyl-3-ammonia-1-
propane sulphonate

xiv

$N^+—CH_2-CH_2-COO^-$ Dodecyl-N-betaine

xiii

Cholate
$3\alpha,7\alpha, 12\alpha$-Trihydroxy-$5\beta$-cholan-24-oate

xv

Decanesulphonate

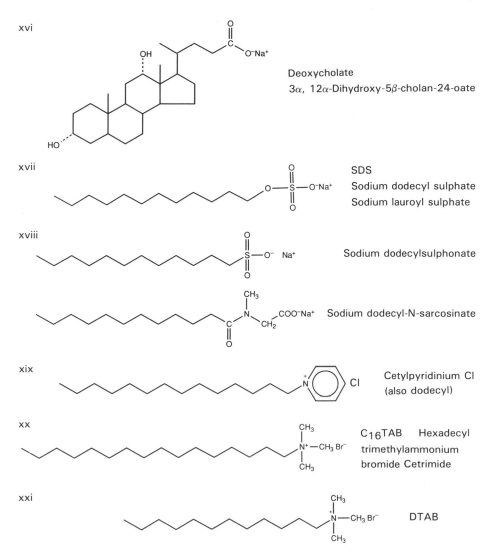

xvi

Deoxycholate
3α, 12α-Dihydroxy-5β-cholan-24-oate

xvii

SDS
Sodium dodecyl sulphate
Sodium lauroyl sulphate

xviii

Sodium dodecylsulphonate

Sodium dodecyl-N-sarcosinate

xix

Cetylpyridinium Cl
(also dodecyl)

xx

$C_{16}TAB$ Hexadecyl
trimethylammonium
bromide Cetrimide

xxi

DTAB

Figure 3. Chemical structures of the principal types of detergent.

instead of SDS (44). The mildest detergents, such as Tween, Lubrol, glucopyranosides, digitonin or Nonidet, can preserve both biological structure and activity but their solubilizing and/or disaggregating abilities are often less than complete. This group are usually non-ionic. Only the more resilient proteins withstand the charged detergents, which therefore constitute an intermediate class. All these detergents can be used with various resins and apparatus which fractionate the polypeptides on the basis of size, but only the non-ionic species are generally applicable with ion-exchange resins.

(b)   If subsequent removal of detergent is envisaged, such as in reconstitution

Table 1. Detergent properties.

Detergent	Charge character	Mol. wt (monomer)	CMC (% w/v)	Mol. wt micelle	Readily dialysable	Comments	Source[a]
Digitonin	Non-ionic	1229		70 000	−	Solubility increased by warming	SCW
α-D-glucopyranoside, octyl	Non-ionic	292	0.29		+	Pure preparations (e.g. Wako) exhibit best solubility and solubilizing properties	SC
β-D-Glucopyranoside	Non-ionic					Hexyl, heptyl and nonyl glucosides also available	
e.g. n-Decyl		320	0.07		−		SC
n-Dodecyl		348	0.0066		−		SC
n-Octyl		292	0.73	8000	+		ACFS
Octyl-β-D-thioglucopyranoside	Non-ionic	308	0.28		+	May be superior to glucosides in some respects	SC
β-D-Maltoside, dodecyl	Non-ionic	511	0.08	50 000	−		SC
N-Methyl glucamides							
Heptanoyl (MEGA-7)		307			+	Mild, non-denaturing; other derivatives can be readily produced	CS
Octanoyl (MEGA-8)		321			+		CS
Nonanoyl (MEGA-9)		335	0.80		+		CS
Decanoyl (MEGA-10)		349	0.80		+		CS
Polyoxyethylene alcohols	Non-ionic					The Brij series also fall into this category and have broadly similar characteristics. So too do the emulphogens which are based on the isoalcohols	
e.g. $C_{11}E_8$		542	0.005	65 000	−		SC
LuBrol PX		582	0.006	64 000	−		SC
Polyoxyethylene octyl phenols	Non-ionic					Polyoxyethylenes prone to oxidation, especially if heavy metal ions are present. Protect with 0.2 mole butylated hydroxytoluene. Absorb strongly at 280 nm, but non-absorbing reduced versions are available from Aldrich	
e.g. Nonidet P-40		603	0.023		−		
Triton X-100		625	0.015	90 000	−		ACFS
Polyoxyethylene sorbitol esters	Non-ionic					Tweens 20, 40 and 60 have CMCs of 0.006, 0.003 and 0.003 (% w/v) respectively	
e.g. Tween 80		1310	0.0015	76 000	−		AFS

CHAPS	Zwitterionic	615	0.49	6150	+	Stored desiccated at RT or 4°C, since hygroscopic	ACFS
CHAPSO	Zwitterionic	631	0.50	6310	+		
Deoxy BIGCHAP	Zwitterionic	878	0.25	8800	+	Reduced interference in ion-exchange chromatography	CS
C16 Lyso PC	Zwitterionic	495	0.0004	92 000	–		CFS
Zwittergent 3-14	Zwitterionic	364	0.011	30 110	–	Range of compounds 3-6 to 3-16 of increasing hydrophobic character. Part of class known as sulphobetaines	C
Cholate, Na	Anionic	430	0.60	4300	+	Also available as the glyco- and tauro-cholates	ACFS
1-Decanesulphonate, Na	Anionic	244	0.80		+	Also pentane and octane sulphonates	S
Deoxycholate, Na	Anionic	414	0.21	4200	+	Also available as the glyco- and tauro-cholates	ACFS
Dodecyl sulphate, Na	Anionic	288	0.23	18 000	partly	K salt insoluble. Strong denaturant but non-denaturing ether sulphates can be synthesized (42)	ACFS
Cetylpyridinium Cl	Cationic	358	0.25		+	Form insoluble complexes with SDS	S
Trimethylammonium bromides	Cationic						
e.g. Cetyl (CTAB)	Cationic	365	0.04	62 000	–	Not readily soluble at 5% w/v in H_2O at room temperature. Strong denaturant	AFS
Dodecyl (DTAB)	Cationic	308	0.43		+		SE
Myristyl (MTAB)	Cationic	336	0.15	19 000	partly		S

[a]A, Aldrich; C, Calbiochem; E, Eastman Kodak; F, Fluka; S, Sigma; W, Wako. Catalogues often contain useful lists of references.

experiments, then it is worthwhile bearing in mind that the higher the CMC, the more easily a detergent will be removed by simple dialysis. Resorting to a number of other procedures to replace detergent with lipid can reduce experimental efficacy and tax protein stability.

(c)   Protein purification often follows extraction from the membrane, and in many circumstances use will be made, during the isolation scheme, of the charge characteristics of the protein. These can often be markedly altered by the use of charged detergents, which may or may not aid purification. In a similar vein, the use of sugar-containing detergents can severely limit the choice of lectin employed in affinity chromatography systems.

(d)   The chemical nature of the detergent should be considered at the outset. If, for example, protein detection or determination will rely on colorimetric assays, then many detergents such as those with amino groups will interfere. Again, if polyacrylamide gel electrophoresis in SDS is to be extensively used, then it is best to avoid the trimethylammonium bromide series of detergents which form insoluble complexes with SDS.

(ii) *Solubilization criteria.* Treatment of membranes with detergent usually results in fragmentation of the structure, but that does not necessarily mean that the protein of interest has been liberated as a monomolecular species. In some instances, microvesicles may be generated or an oligomeric state may be preserved. The cytoskeletal microvillar core, for example, can be recovered still largely undissociated under conditions where the bilayer has been completely dissolved. Therefore, the criteria used to assess solubilization should be carefully considered. Centrifugation at 100 000 g for 1 h will sediment cytoskeletal complexes, and even small vesicles, provided high-density media such as sucrose or glycerol are not employed. Material present in the resultant supernatant fraction is often then described as 'solubilized', but such a description may disguise a range of molecular species. A slightly more definitive, but also more time-consuming, approach is to use density-gradient centrifugation (see Chapters 1 and 2 of ref. 1) or gel filtration on large-porosity resins such as the Sepharoses. Both of these techniques can also give some estimate of molecular weight, and it is recommended that they be used for characterization purposes subsequent to rapid initial centrifugation. The density gradient method for the determination of molecular weight has been refined to accommodate bound detergent (45). Of the other techniques that have been employed to assess solubility, electron microscopy is inconvenient and problematic, turbidity measurements can be misleading, and ^{31}P-NMR has restricted availability.

(iii) *Protein assays.* The presence of phospholipid and detergent, especially the latter, can seriously interfere with protein estimations. Amino acid analysis is the most reliable, but requires significant amounts of protein and relies on the availability of suitable major pieces of equipment. The manual ninhydrin method very successfully copes with most contaminants, while for ease of operation the fluorescamine micro-Lowry, Bradford (Bio-Rad) and BCA (Pierce) methods are very useful (see ref. 46).

(a) *Fluorescamine*

    1. Add 0.2 M sodium borate, pH 9.0−9.5 (proteins), or 8.0−9.0 (peptides) to

the sample to give a volume of about 2 ml and a buffer concentration of over 0.1 M.

2. Add 0.5 ml of 0.2 mg ml^{-1} fluorescamine in acetone; mix thoroughly.
3. Leave for 10 min at room temperature before reading with excitation and emission wavelengths of 390 and 480 nm, respectively.
4. Sensitivity can be increased by hydrolysing protein/peptide with NaOH (see manual ninhydrin below), neutralizing with acetic acid and adding sodium borate buffer to pH 9.0.

(b) *Micro-Lowry A* (adapted from ref. 47)

1. To 200 μl of the protein solution in detergent, add 1 ml of fresh Lowry reagent (1 ml each of 2.0% sodium potassium tartrate and 1% copper sulphate into 100 ml of 2% sodium carbonate in 0.1 M NaOH), mix thoroughly and allow to stand for 10 min.
2. Add 10% SDS to a final level of 1%; mix thoroughly.
3. Add 100 μl of 1:2 diluted Folin−Ciocalteau reagent, mix immediately.
4. Read at 660 nm after 30 min and within 2 h.

(c) *Micro-Lowry B* (adapted from ref. 48). Since some detergents (such as DTAB) are not soluble in SDS, an alternative procedure involving precipitation of protein can be employed.

1. Add 30 μl of 1% sodium deoxycholate to the protein solution (25 μg in 5−200 μl).
2. Add 1 ml of cold 12% trichloroacetic acid.
3. Incubate for 10 min. Spin at 1000 g for 20 min.
4. Add 1 ml of Lowry reagent to the pellet and solubilize (10−15 min).
5. Read at 660 nm after 30 min.

(d) *Manual ninhydrin* (adapted from ref. 49)

1. Dry down the sample in polypropylene tubes.
2. Add 0.15 ml of 9 M NaOH and incubate for 90 min at 110°C.
3. Carefully add 0.25 ml of glacial acetic acid while samples are still warm.
4. Add 0.5 ml of fresh ninhydrin reagent (2 g ninhydrin, 40 mg stannous chloride in 75 ml 2-methoxyethanol and 25 ml 4 M sodium acetate pH 5.5, bubbled with nitrogen) and incubate at 100°C for 15−20 min.
5. Cool, add 2.5 ml of ethanol and read at 570 nm.

(iv) *Solubilization conditions.* Since different membranes and proteins respond to different detergents in various ways, it is not possible to be definitive about the exact conditions which should be employed for solubilization. However, attempts have been made to relate the amount of detergent required for complete solubilization to its CMC and the phospholipid content of the membrane (Chapter 5 of ref. 1; 41,50). Operationally, the best strategy is first to wash the membranes in various salt- and EDTA-buffered solutions to remove peripheral proteins, and then carry out pilot studies, employing a range of detergent concentrations (0.01−5% w/v) in a suitable buffer, to assess the behaviour of the protein generally and the particular component of interest, under the different conditions. Varying the salt content (up to 1 M NaCl) of the detergent solution can also prove beneficial. One then chooses a set of conditions or procedure which generates the highest specific activity of the protein under study. It may be, for example, that the protein of interest is released from the membrane at low detergent concentrations when the rest of the membrane is still relatively intact. Alternatively, the specific protein may require high levels of detergent for solubilization, in which case much other material can first be removed at lower detergent concentrations.

At this stage great care should be exercised to ensure preservation of protein activity. There are few guidelines as to whether any specific protein will be stable in any particular detergent; the only advice that can be given is that the non-ionic detergents have, in general, the least destructive effects. After careful trials with various detergents and conditions of pH and salt concentration, the single most important precaution to take is the addition of protease inhibitors such as PMSF, tetrathionate and phenanthroline, even if they were included during membrane preparation. Disruption of the membrane by detergents often appears to be accompanied by protease activation.

There are several instances where protein activity is lost whenever its normal lipid environment is replaced with detergent (see Chapter 5 of ref. 1). The more effective a detergent is in stripping away residual phospholipid, the greater is the potential for protein instability. Apart from the inclusion of protease inhibitors, the addition of two further ingredients may be helpful. The first of these is $10-50\%$ glycerol, which has the ability to preserve activity not only on solubilization, but also during subsequent fractionation and storage. With high levels of glycerol, centrifugation should be more vigorous (e.g. up to 200 000 g) to ensure that any microvesicles are pelleted through the denser medium. Specific ligands or substrates constitute the second group, and it has occasionally been found that by locking the protein into a defined conformation, its activity can be protected. The addition of agonists, such as isoproterenol, but not antagonists, for example, significantly increases the proportion of active β-adrenergic receptor recovered on solubilization with deoxycholate (51).

Proteins are at their most vulnerable during purification, for it is during these stages that the last vestiges of membrane lipid are removed. On occasion, this removal is accompanied by the loss of biological activity suggesting that some kind of lipid environment is essential. Although the theoretical reasons for this have not been adequately elucidated, it has often been reported that the inclusion of $0.1-5$ mg ml^{-1} exogenous lipid (cholesterol and/or phospholipid) in the fractionation, extraction (52) or buffers (53) does help to preserve protein activity. The exact nature of the lipid can be important, and it may be safest in the first instance to use a mixture which bears some similarity to that in the original membrane.

3. PROTEIN PURIFICATION

As has already been stated, proteins can generally be fractionated on the basis of size, of charge or hydrophobicity, and of specificity. Although this is also true for integral membrane proteins, their hydrophobic nature and the special conditions required for solubilization often impose critical provisos. Peripheral membrane proteins, on the other hand, can be subjected to the procedures for dealing with water-soluble polypeptides described in the companion volume, *Protein Purification Methods: A Practical Approach*. This discussion will deal largely with their application to integral membrane proteins.

3.1 Size-dependent separation

3.1.1 *Organic-acid solvents*

Most carbohydrate-based 'soft' resins (Sepharoses, Sephadexes, Superoses) are not suitable for use with pure or acidified organic solvents. Under such conditions, they do not 'swell' satisfactorily, their fractionating powers are very curtailed and they can strongly adsorb protein. In contrast, synthetic polymer-based resins (such as the hydroxylated polyethers) and hydroxypropylated dextrans (Sephadexes LH20 and LH60) can be used under these conditions in conventional or HPLC−FPLC systems. Silica-based supports function adequately in organic mixtures but degrade slowly in acidic solvents. Polyacrylamides perform well in aqueous acid solvents (even up to 70% formic acid), but are less satisfactory in non-aqueous conditions. The molecular weight range of the proteins to be fractionated will largely determine which pore size of resin (see ref. 54) is most appropriate. In pure organic solvents, the polypeptide may retain much of its native shape, but in FE, FACE or acidified conditions, partial or complete denaturation should be anticipated. The choice of solvent, therefore, will also have an influence on the pore size to be used. It is important to appreciate that all components in contact with organic solvents should be constructed of glass or Teflon. Where highly acidic conditions are being employed, it is often advisable to keep run times and temperatures to a minimum and to wash the resins in milder solvents after use.

3.1.2 *Detergents and denaturants*

Gel filtration in aqueous detergents is often a very useful initial fractionation step. Virtually all resins can be successfully used with virtually all detergents and denaturants (urea, guanidine-HCl). There are a number of points to bear in mind when adopting this approach.

(a)   The column buffer should contain detergent near or above its CMC. This helps ensure that the protein does not aggregate, precipitate or lose its activity during chromatography.

(b)   The association of detergent and protein gives rise to a molecular species of greater molecular weight than the protein alone. The increase in apparent molecular weight is particularly pronounced with denaturing detergents such as SDS, or with proteins which associate with the detergent micelle via say a single intramembranous segment. With mild detergents, oligomeric assemblies present *in situ* in the bilayer may still be preserved when in solution. Some proteins may also aggregate. For these reasons, resins with higher porosities (such as Sepharoses), tend to give more successful fractionation.

(c)   The lipids in the membrane form mixed micelles with detergent. When mild detergents are used, some protein−lipid interactions may survive the solubilization protocol.

(d)   In a few cases, protein can become irreversibly adsorbed to the resin, particularly
        the dextran-based supports. A remedy is often found by changing to more inert
        materials such as the polyacrylamides.

3.2 Charge-dependent separation

3.2.1 *Electrophoresis*

Although often considered somewhat traditional and dated, there are many applications
of electrophoresis on paper, silica thin layers, agarose and polyacrylamide gels which
can be powerfully exploited to yield pure protein or peptide in sufficient quantities for
some types of work, such as microsequencing. The combinations of organic solvents
and acids often employed with paper and silica electrophoresis, for example, can render
peptides in particular sufficiently soluble to effect reasonable fractionation. The reduced
mobilities of whole proteins makes this a less successful approach. However, two
applications of considerable help with integral membrane proteins are polyacrylamide
gel electrophoresis and charge-shift electrophoresis.

(i) *Polyacrylamide gel electrophoresis.* This technique is fully described in ref. 55, but
there are a number of small points which are worth bearing in mind with integral
membrane proteins. Systems have been described whereby electrophoresis can be carried
out in non-denaturing detergents such as Triton X-100 (56), or deoxycholate (57).
Although poor resolution, aggregation and 'smearing' are frequent problems, there are,
nevertheless, some successful applications. One system which holds promise is the use
of the oligooxyethylene dodecyl sulphates, which are not nearly as profoundly denaturing
as SDS, but which bind to membrane proteins in sufficient quantity to induce good
solubility, limited aggregation and good mobility on electrophoresis (44)—all features
which cause problems with the non-ionic detergents.

   Much more frequently, the denaturing sodium and lithium dodecyl sulphates find
favour because of their powerful solubilizing properties and the good resolution that
can be obtained on electrophoresis. These detergents are generally added to at least
1 % or twice the dry weight of the membrane, whichever is the greater. Reducing agents
($1-5$ % 2-mercaptoethanol or, better, 2 mM dithiothreitol) are not essential. The mixture
can then be subjected to heat treatment (60 min at $20-30$°C or $3-5$ min at 100°C).
Although boiling the sample will undoubtedly destroy the vast majority of proteases
still active in SDS, there is the attendant risk of inducing irreversible aggregation of
the protein such as occurs with receptors of the rhodopsin family. In cases such as these,
an equally successful approach would be the addition of 1 mM PMSF, 10 mM EDTA
and possibly also 10 mM tetrathionate and phenanthroline along with the SDS. Following
electrophoresis the proteins can be visualized by a number of techniques and recovered
from the gel by electroelution or diffusion. These aspects are described in ref. 17.

(ii) *Charge-shift electrophoresis.* Charge-shift electrophoresis ($58-62$) exploits the
property of integral membrane proteins to bind detergent to their exposed hydrophobic
segments. Since the non-denaturing detergents used can be neutral, positively or
negatively charged, they can, on binding to the protein, change its charge status and
hence alter its electrophoretic mobility. Different detergents therefore can induce

different electrophoretic behaviour and thus can be used as an indicator of whether the native protein has substantial exposed hydrophobic regions (and hence as to whether it falls into the class known as integral membrane proteins). The procedure described below is relatively straightforward and can be combined with immunoelectrophoresis. The biggest single obstacle is the behaviour of the protein in various detergents, that is, whether it remains as a monodisperse entity, or is prone to aggregation.

(a) Prepare gels on glass plates (10 or 20 × 20 cm) using 1% agarose dissolved in 37.5 mM Tris, 100 mM glycine, pH 8.7, containing 0.5−1% detergent. Neutral detergents include Triton X-100, octyl glucoside, lubrol, emulphogene BC720 (most successful) and digitonin; positively charged detergents include CTAB (most successful) and DTAB; and the most successful negatively charged detergent is deoxycholate. Charged detergents are usually employed as mixtures (1:1−1:10) with the neutral detergent.

(b) Place gels in a flat bed electrophoresis chamber (water-cooled if possible) containing the corresponding buffer. Connect gel and buffer by paper wicks.

(c) Pre-electrophorese for 15−20 min at 4−5 V cm^{-1}.

(d) Apply samples (10−15 μl containing about 10−20 μg protein in up to 5% single or mixed detergent) into slits (1 × 10 mm) cut into the agarose. The origin should be approximately equidistant from the wicks, unless the protein has pronounced charge characteristics.

(e) Electrophorese for 2−3 h at 4−5 V cm^{-1}.

(f) Remove gel, dry with a warm air fan (may need to fix and press), stain for protein with 0.1% Coomassie blue, 40% methanol, 7% acetic acid for 1−3 h, then destain in the same solution minus the dye.

(g) Proteins which bind DOC or CTAB should exhibit anodic and cathodic shifts respectively, compared with their positions in a neutral detergent such as Triton.

This method is particularly powerful when combined with crossed immunoelectrophoresis to give crossed immuno-charge-shift electrophoresis. In this technique, electrophoresis in neutral or charged detergent is performed along one edge of square glass plates as described above. The bulk of the agarose above the sample is then removed and replaced with agarose which contains antibody to the specific protein, along with the neutral detergent alone. Second-dimension electrophoresis in the manner conventional for crossed immunoelectrophoresis then gives precipitation patterns in which charge-shifts are particularly easy to detect.

A further extension of this approach can be used to detect, and possibly subsequently identify, protein receptors for ligands (62). Solubilized membranes plus bound ligand are subjected to electrophoresis in detergent in the first dimension, followed by immunoelectrophoresis in the second dimension. Antibody against the ligand allows the detection of the receptor−ligand complex which migrates in a position different from that of the free ligand. This approach can also be used for non-protein ligands provided they can be coupled in active form to a carrier protein with antigenic properties.

(iii) *Isoelectric focusing.* For the recovery of small amounts of protein in very pure form, isoelectric focusing, particularly when coupled with electroelution, presents a powerful practical tool. Membrane proteins exhibit a wide range of isoelectric points and, provided they can be obtained in a non-aggregated state, isoelectric focusing in detergent can be very successful, to the point where differently phosphorylated versions of the same protein can be resolved (63). The procedures for integral membrane proteins are little different to those outlined in the companion volume (17) for water-soluble proteins, except for the inclusion of detergents, which should usually be of the non-ionic type (zwitterionic detergents can also be used) to guard against overwhelming the native charge on the protein. When the pI is 7 or below, it is often advisable to use non-equilibrium pH gradient electrophoresis (Chapter 6 of ref. 1; 64).

3.2.2 *Ion-exchange chromatography*

All the usual ion-exchange systems used in protein purification can be recruited for the purification of integral membrane proteins. As with isoelectric focusing, detergents are required at all times, and the choice of detergent is again governed by the need to preserve both the native structure and charge characteristics of the protein. Non-ionic detergents are usually preferred, but zwitterionic species can also be used. The relatively new detergent BIGCHAP (see *Table 1*), a non-ionic member of the CHAP series, may provide the best of both worlds. DEAE- and CM-derivatized supports and hydroxyapatite have been successfully employed with membrane proteins (65), but the relative efficacy of these resins largely rests with the nature of the protein. The object is often to devise conditions whereby the polypeptide of interest forms a reversible interaction with the functional groups on the support. In general, this is best achieved in those cases where a large proportion of the mass of the protein is exposed to the aqueous phase rather than smothered with detergent. The detergent itself is usually added at, or above, its CMC, and in some cases it has been found necessary to include exogenous lipid in the buffers (53). But for the inclusion of detergent, the operating conditions for these resins are as described in the companion volume (66). These comments all apply to water-based buffers, but there have been applications of DEAE-(34) and CM-(33) cellulose with organic mixtures such as chloroform−methanol. In these systems the ratio of chloroform to methanol and the water content of the eluting buffer have been varied. Such applications are limited to proteins soluble in such conditions, and are only rarely appropriate and successful.

3.2.3 *Chromatofocusing*

One of the more recent and promising developments in the approaches to purification has been the technique of chromatofocusing (66). The supports used in this technique are substantially resistant to degradation between pH 3 and 12, but are more successful in the pH range 5−9. They can be used with neutral organic solvents, with denaturants such as 8 M urea or 6 M guanidine hydrochloride, and with uncharged or zwitterionic detergents. The method is rapid and efficient, if the protein remains soluble and disaggregated. We have found it particularly useful with rhodopsin, using detergents such as digitonin and sodium monolaurate at concentrations above their CMCs.

3.3 **Separation based on hydrophobicity**

3.3.1 *Hydrophobic interaction chromatography (HIC)*

Most proteins have on their accessible surfaces small hydrophobic patches through which they can be induced to associate with hydrophobic substituents on insoluble supports. This forms the basis of a useful purification approach called hydrophobic interaction chromatography which is described fully in ref. 66 (reverse-phase chromatography is a variant on this theme and is described separately below). The resins are generally dextran-based (Sepharoses or Superoses) and are derivatized with uncharged moieties (hexyl, octyl, phenyl) through which interaction with the protein occurs. Although generally a successful technique, this form of hydrophobic interaction chromatography has only limited value for integral membrane proteins. The essential presence of detergents effectively abolishes any meaningful hydrophobic interaction, as will any substantial amount of organic solvent.

3.3.2 *Reverse-phase chromatography (RPC)*

This variant of the technique particularly developed for HPLC and FLPC systems is somewhat more successful than HIC with membrane proteins and peptides. The resins, their character and their uses are described in ref. 66. Most integral proteins or peptides generally need to be applied in a small volume of solvent, such as $50-70\%$ formic acid in water, in order to maintain their solubility. For this reason, RPC of membrane proteins is rarely consistent with preservation of biological activity. The column may then be developed with conventional systems such as $0.01-0.1\%$ trifluoroacetic or pentafluorobutyric acid containing increasing concentrations of an organic solvent to generate a gradient of decreasing polarity. The polarity of these solvents decreases in the order methanol, ethanol, acetonitrile, 1-propanol, 2-propanol and butanol. Unfortunately, those proteins or peptides with pronounced hydrophobic character are often not recovered in good yield using this type of elution system. We have had some success with more aggressive gradients, such as 5% formic acid in water to 5% formic acid in ethanol (67) or, with particularly hydrophobic peptides, 50% formic acid in water to 50% formic acid in isopropanol or ethanol (see Chapter 6, ref. 1). Only FPLC and robust HPLC systems can tolerate these more destructive conditions. Other solvent mixtures which have been useful include (i) 0.24 M potassium acetate pH 6.5 in chloroform:methanol (1:2 v/v) containing 8% H_2O to 0.24 M potassium acetate in methanol:water (3:1 v/v) (68); (ii) various chloroform:methanol mixtures (69); and (iii) gradients of 12 mM HCl in 1% ethanol:1-butanol (4:1 v/v) to 12 mM HCl in 60% ethanol:1-butanol (4:1 v/v) (70). Additives such as monomethoxyethylene glycol or dimethylsulphoxide are also reported to increase recovery, but their applicability to hydrophobic peptides is limited in view of the high concentrations of non-polar eluant required. Finally, silica gel has been employed to purify proteolipid proteins using a benzene−ethanol gradient of 95:5 to 44:56 (71).

3.4 **Affinity chromatography**

For a variety of reasons articulated in the companion volume (72) as well as in separate texts (Chapter 6 of ref. 1; 73), affinity chromatography presents one of the most rapid, efficient and convenient means of generating pure protein. It has been a particularly valuable technique in the field of membrane proteins where stability can be a problem and the amounts of available protein are severely limiting. If antibodies can be raised against the protein, immunoadsorbents can be a very powerful tool (Chapter 3 of ref. 1; 74).

The principal factor to bear in mind with membrane proteins is the possible large size of the protein−detergent complex. For this reason it is often advisable to employ gels with high porosity and/or which possess extension arms (72). Both these features help to reduce the effects of steric hindrance.

Any ligand with an affinity constant for the protein of 10^{-7} M or better can be used to effect successful purification. Since many membrane proteins are glycosylated, carbohydrate-binding lectins have proved to be very useful (see ref. 75). They can be coupled readily to a variety of supports and are thoroughly stable in non-denaturing detergents. They possess different sugar specificities (76) and can thus be employed to differentiate between glycoproteins (77,78). Detergents which contain sugar moieties should be tested to ensure they do not bind to the active site of the lectin to be used. It is also relevant to point out that the carbohydrate side-chains of many proteins may be heterogeneous due, for example, to glycosidase action. In these cases, the same protein may appear to exhibit schizophrenic behaviour in its binding to a particular lectin (77). Where the glycoprotein of interest contains a terminal sialic acid residue, most of the other membrane glycoproteins can be removed from the mixture by using a combination of lectins of different specificity (for example *Lens culinaris* phytohaemagglutinin and wheat-germ agglutinin). The unbound sialic-acid-containing fraction can then be treated with neuraminidase to remove this residue, thereby generating a new terminal sugar moiety which is now recognized by a lectin-affinity support (79).

Glycoproteins can also be bound non-specifically by boronate resins (72,73) and recovered by elution with buffers containing sorbitol, citrate or acetate, or at low pH (25 mM HCl, 100 mM formic acid). This approach is relatively new, and although successful with glycosylated haemoglobins, it has not yet been reported for membrane proteins.

4. CONCLUSION

The purification of membrane proteins still presents the research worker with major problems, not the least of which is the need to preserve biological activity when samples are in soluble form. The number and variety of detergents increase daily, together with enhanced purity and preserving power, but we are still far from the position of obtaining an effective detergent with the ability to substitute functionally for the phospholipid environment. There also seems to be great scope for the development of new chromatography resins and solvent systems for coping with this hydrophobic class of protein.

5. REFERENCES

1. Findlay,J.B.C. and Evans,W.H. (eds) (1987) *Biological Membranes. A Practical Approach.* IRL Press, Oxford.
2. Bordier.C. (1981) *J. Biol. Chem.,* **256**, 1604.
3. Clemetson,K.J., Bienz,D., Zahno,M.-L. and Luscher,E.F. (1984) *Biochim. Biophys. Acta,* **778**, 463.
4. Mahar,P.A. and Singer,S.J. (1985) *Proc. Natl. Acad. Sci. USA,* **82**, 959.
5. Marchesi,V.T. and Steers,E., Jr (1968) *Science,* **159**, 203.
6. Fairbanks,G., Steck,T.L. and Wallach,D.F.H. (1971) *Biochemistry,* **10**, 2606.
7. Tanner,M.J.A. and Gray,W.R. (1971) *Biochem. J.,* **125**, 1109.
8. Beynon,R.J. (1989) In *Protein Purification Methods: A Practical Approach.* Harris,E.L.V. and Angal,S. (eds), IRL Press, Oxford, p. 40.
9. Steck,T.L. (1972) *Biochim. Biophys. Acta,* **225**, 553.

10. Carter,J.R., Jr (1973) *Biochemistry*, **12**, 171.
11. Schiechl,H. (1973) *Biochim. Biophys. Acta*, **307**, 65.
12. Steck,T.L. and Yu,J. (1973) *J. Supramol. Struct.*, **1**, 221.
13. Laperche,Y., Bulle,F., Aissani,T., Chobert,M.-N., Aggerbeck,M., Hanoune,J. and Guellaen,G. (1986) *Proc. Natl. Acad. Sci. USA*, **83**, 937.
14. Wu,H.C. and Tokunaga,M.L. (1986) *Curr. Top. Microbiol. Immunol.*, **125**, 127.
15. Low,M.G. (1987) *Biochem. J.*, **244**, 1.
16. Davitz,M.A., Hereld,D., Shak,S., Krakow,J., Englund,P.T. and Nussenzweig,V. (1987) *Science*, **238**, 81.
17. Dunn,M.J. (1989) In *Protein Purification Methods: A Practical Approach.* Harris,E.L.V. and Angal,S. (eds), IRL Press, Oxford, p. 18.
18. Mali,A.-S. and Low,M.G. (1986) *Biochem. J.*, **240**, 51.
19. Ulrich,A., Bell,J.R., Chen,E.Y., Herrera,R., Petruzzelli,L.M., Dull,T.J., Gray,A., Coussens,L., Liao,Y.C., Tsubokawa,M., Mason,A., Seeburg,P.M., Greenfield,C., Rosen,O.M. and Ramachandran,J. (1985) *Nature*, **313**, 756.
20. Sudhof,T.C., Goldstein,J.L., Brown,M.S. and Russell,D.W. (1985) *Science*, **228**, 815.
21. Findlay,J.B.C. and Pappin,D.C. (1986) *Biochem. J.*, **238**, 625.
22. Folch,J., Lees,J. and Slone-Stanley,G. (1957) *J. Biol. Chem.*, **226**, 497.
23. DeRobertis,E. (1975) *Rev. Physiol. Biochem. Pharmacol.*, **73**, 11.
24. Altendorf,K., Lukas,M., Lohl,B., Muller,C. and Sandermann,H. (1977) *J. Supramol. Struct.*, **6**, 229.
25. Phizackerly,P.J.R., Town,M. and Newman,G.E. (1979) *Biochem. J.*, **183**, 731.
26. DeRobertis,E., Fiszerde Plazas,S., Llorente de Carlin,C., Aguilar,J. and Schlieper,P. (1978) *Adv. Pharmacol. Ther.*, **1**, 235.
27. Alterdorf,K. (1977) *FEBS Lett.*, **73**, 271.
28. Boyan-Salyers,B., Vogel,J., Riggan,L., Summers,F. and Howell,R. (1978) *Metab. Bone Dis. Rel. Res.*, **1**, 143.
29. Criado,H., Aquilar,J. and DeRobertis,E. (1980) *Anal. Biochem.*, **103**, 289.
30. Brett,M. and Findlay,J.B.C. (1983) *Biochem. J.*, **211**, 661.
31. Pappin,D.J.C. and Findlay,J.B.C. (1984) *Biochem. J.*, **217**, 605.
32. Findlay,J.B.C., Brett,M. and Pappin,D.J.C. (1981) *Nature*, **293**, 314.
33. Siebald,W. and Wachter,E. (1980) *FEBS Lett.*, **122**, 307.
34. Fillingame,R. (1976) *J. Biol. Chem.*, **251**, 6630.
35. Tandy,N.E., Dilley,R.A., Hesmondson,H.A. and Bhatnagar,D. (1982) *J. Biol. Chem.*, **257**, 4301.
36. Berger,G.E., Anderegg,R.J., Herlihy,W.C., Gray,C.P., Biemann,K. and Khorana,H.G. (1979) *Proc. Natl. Acad. Sci. USA*, **76**, 227.
37. Marchesi,V.T. and Andrews,E.P. (1971) *Science*, **174**, 1247.
38. Helenius,A. and Simons,K. (1975) *Biochim. Biophys. Acta*, **415**, 29.
39. Reynolds,J.A. (1982) In *Lipid—Protein Interactions.* Jost,P.C. and Griffith,O.H. (eds), John Wiley, New York, Vol. 2, p. 193.
40. Lichtenberg,D., Robson,R.J. and Dennis,E.A. (1983) *Biochim. Biophys. Acta*, **737**, 285.
41. Womack,M.D., Kendall,D.A. and Macdonald,R.C. (1983) *Biochim. Biophys. Acta*, **733**, 210.
42. Elworthy,P.H., Florence,A.T. and Macfarlane,C.B. (1968) *Solubilization by Surface Active Agents.* Chapman and Hall, London.
43. Mukerjee,P. (1967) *Adv. Coll. Int. Sci.*, **1**, 241.
44. Koide,M., Fukuda,M., Ohbu,K., Watanabe,Y., Hayashi,Y. and Takagi,T. (1987) *Anal. Biochem.*, **164**, 150.
45. Clarke,S. (1975) *J. Biol. Chem.*, **250**, 5459.
46. Dunn,M.J. (1989) In *Protein Purification Methods: A Practical Approach.* Harris,E.L.V. and Angal,S. (eds), IRL Press, Oxford, p. 10.
47. Markwell,M.A.K., Haas,S.M., Bieber,L.L. and Tolbert,N.E. (1978) *Anal. Biochem.*, **87**, 206.
48. Bensadoun,A. and Weinstein,D. (1976) *Anal. Biochem.*, **70**, 241.
49. Hirs,C.H.W. (1967) In *Methods in Enzymology.* Hirs,C.H.W. (ed.), Academic Press, New York, Vol. 11, p. 325.
50. Rivnay,B. and Metzger,H. (1982) *J. Biol. Chem.*, **257**, 12800.
51. Nedive,E. and Schramm,N. (1984) *J. Biol. Chem.*, **259**, 5803.
52. Anholt,R., Fredkin,D.R., Deerinck,T., Ellisman,M., Montal,M. and Lindstrom,J. (1982) *J. Biol. Chem.*, **257**, 7122.
53. Bogonez,E. and Koshland,D.E., Jr (1985) *Proc. Natl. Acad. Sci. USA*, **82**, 4891.
54. Prenata,A. (1989) In *Protein Purification Methods: A Practical Approach.* Harris,E.L.V. and Angal,S. (eds), IRL Press, Oxford, p. 293.
55. Hames,B.D. and Rickwood,D. (eds) (1981) *Gel Electrophoresis of Proteins: A Practical Approach.* IRL Press, Oxford.

56. Dewald,B., Dulaney,J.T. and Touster,O. (1974) In *Methods in Enzymology*. Fleischer,S. and Packer,L. (eds), Academic Press, New York, Vol. 32, p. 82.
57. Dulaney,J.T. and Touster,O. (1970) *Biochim. Biophys. Acta,* **196**, 490.
58. Helenius,A. and Simons,K. (1977) *Proc. Natl. Acad. Sci. USA,* **74**, 529.
59. Norrild,B., Bjerrum,D.J. and Vestergaard,P.F. (1977) *Anal. Biochem.,* **81**, 432.
60. Bjerrum,O.J. (ed.) (1983) *Immunoelectrophoretic Analysis of Membrane Proteins*. Elsevier, Amsterdam.
61. Booth,A.G., Hubbard,L.M.L. and Kenny,A.J. (1979) *Biochem. J.,* **179**, 397.
62. Booth,C.M. and Booth,A.G. (1982) *Placenta,* **3**, 57.
63. Aton,B.R., Litman,B.J. and Jackson,M.L. (1984) *Biochemistry,* **23**, 1737.
64. O'Farrell,P.Z., Goodman,H.M. and O'Farrell,P.H. (1977) *Cell,* **12**, 1133.
65. Sigel,E., Stephenson,F.A., Mamalaki,C. and Barnard,E.A. (1983) *J. Biol. Chem.,* **258**, 6965.
66. Roe,S. (1989) In *Protein Purification Methods: A Practical Approach*. Harris,E.L.V. and Angal,S. (eds), IRL Press, Oxford, p. 175.
67. Khorana,H.G., Gerber,G.E., Herlihy,W.C., Anderegg,R.J., Gray,C.P., Nihei,K. and Biemann,K. (1979) *Proc. Natl. Acad. Sci. USA,* **76**, 5046.
68. Blondin,G. (1979) *Biochem. Biophys. Res. Commun.,* **87**, 1087.
69. Blondin,G. (1979) *Biochem. Biophys. Res. Commun.,* **90**, 355.
70. Van der Zee,R., Welling-Webster,S. and Welling,G. (1983) *J. Chromatogr.,* **266**, 577.
71. Macklin,W., Pickart,L. and Woodward,D. (1981) *J. Chromatogr.,* **210**, 174.
72. Angal,S. and Dean,P.D.G. (1989) In *Protein Purification Methods: A Practical Approach*. Harris,E.L.V. and Angal,S. (eds), IRL Press, Oxford, p. 245.
73. Dean,P.D.G., Johnson,W.S. and Middle,F.A. (eds) (1985) *Affinity Chromatography: A Practical Approach*. IRL Press, Oxford.
74. Hill,C.R.H., Thompson,L.G. and Kenney,A.C. (1989) In *Protein Purification Methods: A Practical Approach*. Harris,E.L.V. and Angal,S. (eds), IRL Press, Oxford, p. 282.
75. Sutton,C. (1989) In *Protein Purification Methods: A Practical Approach*. Harris,E.L.V. and Angal,S. (eds), IRL Press, Oxford, p. 268.
76. Goldstein,I.J. and Hayer,C.E. (1978) *Adv. Carbohydr. Chem. Biochem.,* **35**, 127.
77. Findlay,J.B.C. (1974) *J. Biol. Chem.,* **249**, 4398.
78. Kahane,I., Furthmayer,H. and Marchesi,V.T. (1987) *Biochim. Biophys. Acta,* **426**, 464.
79. Carter,W.G. and Sharon,N. (1979) *Arch. Biochem. Biophys.,* **180**, 570.

CHAPTER 5

Purification of proteins for sequencing

J.B.C.FINDLAY

1. INTRODUCTION

With the advent of recombinant DNA techniques, the role of protein sequencing has changed. Due to the speed and ease of sequencing DNA, the complete primary structures of large and/or scarce proteins are usually more easily determined by translation of the cDNA sequence. However, short regions of the protein sequence need to be determined for a large number of applications, e.g. synthesis of oligonucleotide probes for gene cloning, topographic analysis, protein processing and isoenzyme identification. Again, attention is increasingly focused on regions of structural and functional importance in proteins. Very often such regions are the sites of covalent modification. In all such instances, therefore, protein sequencing is the only method which can be used to determine the location of the modified amino acids within the primary structure.

Current methods of protein sequence analysis are quantitative and can therefore be used to assess the purity of a peptide or protein preparation, or to determine subunit stoichiometry of complex multi-subunit systems.

These methods fall into two categories: Edman degradation (1), or mass spectrometry employing Fast Atom Bombardment (FAB) (2). Edman degradation can be performed manually using phenyl isothiocyanate alone or with dimethylaminobenzene isothiocyanate (3) or fluorescein isothiocyanate (4). Quantities of protein or peptide down to below 1 nmole and 20 pmole respectively can be sequenced, with a realistic length of about 20 residues. Alternatively, the Edman degradation can be performed using automated methods where sensitivities of below 10 pmole are readily achieved and 50−80 residues can be obtained from 1 nmole of starting material. In the solid-phase sequencing approach (5), the peptide or protein is covalently linked to a solid support to minimize losses during the sequence reaction and to increase chemical cleanliness. 'Gas-phase' or liquid pulse sequencers (6) have largely superseded the liquid spinning cup sequencers due to their improved sensitivity. With these machines, the peptide or protein is adsorbed non-covalently on to an inert support.

Use of FAB mass spectrometry has some advantages over the Edman degradation. Peptides with blocked N-termini can be sequenced, and, since information can also be obtained on the carbohydrate structure, glycoproteins can potentially be characterized in a single step. In addition, the sequences of mixtures of up to five peptides can be unambiguously determined. However, FAB mass spectrometry cannot be used for sequencing whole proteins directly; shorter fragments must first be generated. It also still lacks some of the exquisite sensitivity achieved with the Edman degradation approach.

This chapter describes some of the precautions necessary when purifying proteins for amino acid sequencing and some techniques frequently used for final purification prior to sequencing. Information on modern techniques used for protein sequencing can be found in ref. 5.

2. TECHNIQUES

2.1 **General precautions**

All the approaches to protein and peptide purification detailed in the companion volume (7) will generate material suitable for manual or automatic sequence analysis. However, a number of points, briefly discussed below, should be borne in mind when preparing samples specifically for protein sequencing. These strictures apply more particularly to the Edman degradation approach to sequencing than to the use of FAB mass spectrometry which had its own set of proscribed contaminants (see ref. 5).

To avoid *in-vitro* blockage of the N-terminus by impurities it is important to employ high-quality reagents (Analar or better) and mild conditions throughout the purification. For example, solutions of urea contain cyanate ions which react with amino groups; high-quality urea contains a lower proportion of these ions. As a further precaution, urea solutions should be made immediately prior to use and purified on a mixed-bed ion-exchange resin (available from BDH and Bio-Rad). The addition of 20 mM methylamine may also help. Specific reagents which are known to react with amino groups should not be used.

Of the small molecules which are undesirable, compounds containing a primary amine are usually the most disadvantageous and often totally undermine sequencing analysis. In solid-phase sequencing, such substances compete very efficiently with proteins and peptides in some of the methods used for attachment to the support. In other forms of automated and manual sequencing, they react with the Edman degradation reagent phenylisothiocyanate, thus giving rise, at the least, to a reduced repetitive yield or more seriously to a significant overlap problem. Sequence overlap is caused by partial reaction during one cycle of the sequencing, thus from this cycle onwards a staggered sequence will be obtained with at least two amino acids being observed at each subsequent cycle. Both events will minimize the length of unambiguous amino acid sequence which can be obtained.

Perhaps the commonest source of amino groups comes from the use of ammonium-containing salts, particularly ammonium bicarbonate; this substance is often thought to be removed completely by lyophilization, but in the author's experience complete removal is never achieved by this method. Other commonly used compounds which should be avoided include Tris, pyridine, glycine, bicine, amino-sugars, polybuffers, ampholytes and most detergents. Although sequencing, especially using the solid-phase methodology, can still be carried out in the presence of some of the latter group of substances, impurities may be present which seriously reduce the efficiency of the sequence procedures. Salts of most kinds should be eliminated if the sample is to be subjected to FAB-MS since they can produce undesirable adducts with glycerol, the carrier usually employed in this technique. Macromolecular contaminants such as phospholipids, carbohydrate and nucleic acid are less frequently encountered, but each presents its own special difficulties which necessitates their removal.

2.2 **Final purification steps**

There are a large number of approaches which can be taken to remove impurities likely to interfere with the various sequencing procedures. These are briefly summarized below. The method chosen will largely depend on the nature and amount of material prepared for sequencing, but attention is drawn particularly to the ease and efficiency of SDS−PAGE [see (v) below].

(i) *Dialysis.* Dialysis against water will in many cases be sufficient. In some instances, and certainly with solid-phase sequencing approaches, the inclusion of 0.1−0.25% SDS may retain the protein in solution without deleteriously affecting sequencing (in fact it can be beneficial during the coupling of protein to resin). If removal of SDS is required, this can usually be achieved by dialysis against 20−30% ethanol or 2-propanol. Dialysis is not recommended with less than 1 nmole of material unless some of the microdialysis systems are available (see Pierce). SDS should be avoided with gas- or liquid-pulse machines.

(ii) *Reverse-phase chromatography.* Sep-Pak cartridges can be helpful and are simple to use. The protein is adsorbed on to the resin, salts and contaminants can then be washed out with water and the protein recovered with acetonitrile or methanol.

Reverse-phase HPLC represents a very effective means of obtaining pure material. Organic solvents (particularly acetonitrile which can be obtained in good purity) in 0.01−0.1% trifluoroacetic acid are particularly good. The samples can then be freeze-dried to remove solvent. (Note that on redissolution the solution could be acidic.) Hydrophobic interaction chromatography can also be very useful for the removal of ampholytes, polybuffers and salts, provided the protein can be eluted in low ionic strength solutions.

(iii) *Precipitation with organic solvents.* Precipitation and washing in 10−20 vols of cold acetone, or 5−10 vols of cold diethyl ether, or ether−ethanol combinations (1:1 to 3:1 v/v), is also a rapid and effective method for cleaning up samples, especially if removal of detergents is required. The use of trichloroacetic acid is not advised.

(iv) *Gel-filtration chromatography.* Desalting using Sephadexes G-10 and G-25 or Biogels P-2 and P-4 in 1−5% acetic or formic acid is an efficient method. A better system, where membrane proteins and/or detergents are involved, which will usually retain these components as monomers in solution, is formic acid−acetic acid−chloroform−methanol (1:1:2:1 v/v) used with Sephadex LH-60 and LH-20 resins (8). On drying samples from these solvents, do *not* allow the material to warm up, otherwise excessive N-terminal formylation will result.

Alternatively, an HPLC gel permeation column, such as the TSK-series (Anachem, Beckman, Pharmacia-LKB) or the GF-series (DuPont) can be used. In general, gel filtration is best avoided unless the protein or peptide is present in nanomolar quantities.

(v) *Polyacrylamide gel electrophoresis.* Finally, a rapid and efficient way of removing small amounts of contaminating material, particularly macromolecules, is to carry out polyacrylamide gel electrophoresis in sodium dodecyl sulphate (SDS) followed by electroelution or electroblotting. To avoid blocking the N-terminus of the protein it is important to use high-quality reagents for making the gel. 0.2 mg ml^{-1} thioglycolate in the running buffer will also help. Electrophoretic-grade reagents (e.g. BDH) should be used, and ideally SDS should be recrystallized from warm-water solutions.

A simple protocol for electroelution requiring no complicated apparatus is given in *Method Table 1*. For use in solid-phase protein sequencers, proteins can also be electro-phoretically transferred directly from the gel on to derivatized glassfibre paper or polyvinylidene difluoride (PVDF) (9,10). PVDF membranes without prior activation can be used for gas-phase machines (11). The transferred proteins are visualized by Coomassie blue or a fluorescent dye, the spots or bands excised and placed directly into the sequencer.

Method Table 1. Electroelution of proteins from SDS−polyacrylamide gels

---

Equipment/reagents

1. Flat-bed gel tank of the type commonly used for DNA separations or an adapted tube gel apparatus.
2. Dialysis tubing, boiled for 15 min in 0.1 M sodium carbonate with 20 mM EDTA, then thoroughly rinsed with distilled water—use Spectrapor for small proteins/peptides.
3. Coomassie stain solution: 0.1% w/v Coomassie brilliant blue R in 50% v/v aqueous methanol, 7% v/v acetic acid. The destain solution is identical, minus dye.

Procedure

1. Stain the gel very briefly with Coomassie blue (5−10 min), then rinse in destain solution until the protein bands become visible (10−15 min). Keep stain and destain times as short as possible for maximum yields.
2. Excise the band of interest. Do not break up or homogenize the gel slice.
3. Fill the flat-bed tank with buffer (25 mM Tris−glycine pH 8.5, 0.1% w/v SDS; 50 mM Tris−acetate, pH 7.8, 0.1% SDS; 0.1 M sodium bicarbonate pH 7.8, 0.1% SDS or 0.1 M sodium phosphate pH 7.8, 0.1% SDS) to a level approximately 1 cm above the platform. 0.1% v/v 2-mercaptoethanol or 2−5 mM dithiothreitol can be added to the buffer if required. For membrane proteins, the SDS concentration of the buffer can be raised to 1% w/v to ensure solubility.
4. Cut a length of dialysis tubing long enough to take the gel slice plus approximately 2 cm at either end. Clip one end with a Mediclip and fill the tubing with buffer from the tank. Place the excised gel slice in the tubing and gently squeeze out most of the liquid before sealing. Position the gel slice on one side of the dialysis bag. Care should be taken to avoid trapping any air bubbles.
5. Place the dialysis bag on to the platform of the electrophoresis tank. The level of buffer should be adjusted so that it just covers the bag.
6. Electrophorese for 3−20 h at 25−100 V constant current (20−150 mA), then reverse the current for approximately 30 s (in order to electrophorese the protein

off the dialysis membrane surface). Longer staining/destaining periods require more extended electroelution times.

7. Remove the gel slice from the bag. Re-stain to check that the protein has been eluted. All small fragments of gel that remain must be removed. It is usually best to decant the solution and centrifuge briefly or filter to remove any small pieces. Re-seal the bag and dialyse the protein solution against at least 5 changes of distilled water (5 litres each) over $2-3$ days at $0-4°C$. The dialysis solutions can contain up to 0.25% w/v SDS to keep the protein in solution, but make the last change against distilled water only. The Coomassie dye stays with the protein throughout and can thus be used as an indicator of the progress of the electroelution.

Modifications

1. The procedure can be used with all types of staining except the silver method which blocks the protein, and with any apparatus based on tube gels where the dialysis bag is fixed on to the bottom of the tube.

2. Potentially the greatest losses are experienced during the fixing, staining and destaining steps. This can be circumvented by staining *before* or *during* electrophoresis. The mobility of the protein is not normally altered. Add Coomassie blue G to the cathode buffer to 25 mg per litre. After electrophoresis, the protein-containing bands can be directly excised and subjected to the electroelution procedure.

2.3 **Peptide production and purification**

For a large number of reasons it is often necessary to generate peptides from a protein for sequencing. If sequence information is required for generating DNA probes for use in gene cloning, peptides containing methionine or tryptophan can be very useful, since these amino acids are encoded by only one triplet of bases. Tryptophan-containing peptides can be detected by absorbance at 280 nm; ideally the eluate from the chromatography is monitored with a dual-channel detector (set at 220 and 280 nm) or a diode-array detector (which measures the absorption spectrum of each peak). Methionine-containing peptides can be detected by amino-acid analysis or by specific labelling (5). Peptides 30 residues or below are required for good sequencing by FAB mass spectrometry. Commonly used methods for generating and purifying such peptides are given below and in ref. 5.

However, by far the most common need is to generate sequence data from proteins that are N-terminally blocked. In such circumstances, the most fruitful approach is to carry out limited proteolysis using enzymes such as endoproteinases Arg-C, Asp-N, Glu-C and Lys-C (*Method Table 2*). Proteolysis can be carried out on electroeluted protein in the presence of up to 1% SDS and the method can be readily amalgamated with SDS−PAGE. Following digestion, the resultant peptides can be separated and recovered by SDS−PAGE. Full experimental details are given in Chapter 2 of ref. 5.

For the generation of smaller peptides, a number of methods are available, but first it is often advisable to reduce the protein and irreversibly modify all cysteine residues.

Method Table 2. Cleavage methods

Agents	Conditions
Endoproteinase Arg-C	100 mM NH_4HCO_3, 20−37°C for up to 8 h. Terminate digestion by freezing, boiling or 10 mM PMSF.
Endoproteinase Asp-N	50 mM Na phosphate pH 8.0, 20−37°C for up to 18 h. Can include up to 1 M urea, 1 M guanidinium−Cl or 0.01% SDS.
Endoproteinase Glu-C	As for Arg-C, up to 0.5% SDS.
Endoproteinase Lys-C	As for Arg-C, up to 0.1% SDS.
Trypsin	As for Arg-C.
Cyanogen bromide (CNBr)	70% HCOOH, up to 1000-fold molar excess over Met, under N_2, in dark at room temperature for up to 24 h. Terminate by rotary evaporation. (**Take care**: CNBr gives rise to toxic products.)

Method Table 3. Reduction and sulphydryl modification

1. Dissolve the protein in up to 1 ml 6 M guanidinium−Cl or 8 M deionized urea in 0.5 M Tris−HCl pH 8.0−8.5, 1 mM EDTA. Up to 1% SDS may also be included to ensure solubility.
2. Add 10 μl 2-mercaptoethanol or DTT to 10 mM and incubate for 1−3 h at 20−35°C in the dark under N_2.
3. Add colourless iodoacetate or iodoacetamide (up to 0.1 ml of 250 mg ml^{-1} in 0.1 M NaOH) or 4-vinylpyridine (to 25 mM in acetonitrile) and incubate in the dark at room temperature under N_2 for 20 min or up to 2 h respectively.
4. Recover protein by precipitation in acetone, dialysis or gel filtration.

This prevents disulphide formation and/or exchange and allows subsequent identification of the cysteine during sequencing (unmodified cysteines are converted to products which are difficult to detect). Reagents commonly used for this include iodoacetic acid and 2-vinylpyridine (*Method Table 3*).

The most successful methods currently used for generating small peptides are (i) chemical cleavage after methionine residues using cyanogen bromide; or (ii) enzymatic cleavage after lysine or arginine using trypsin, or after glutamic acid using *Staphylococcus aureus* V8 endoproteinase Glu-C. A number of other chemical and enzymic methods can also be successful. These are fully described in ref. 5.

The mixture of peptides can be substantially purified by HPLC or FPLC using reverse-phase and/or ion-exchange chromatography. Since the mixtures are normally complex, more than one chromatography method is usually required. Thus, ion-exchange chromatography may be used first, followed by reverse-phase systems. Alternatively reverse-phase chromatography at two different pH values could be tried. Ideally the final (reverse-phase) chromatography should be carried out in a volatile buffer such as 0.1% trifluoroacetic acid and acetonitrile. The peptides can therefore be subsequently dried *in vacuo* using a centrifugal concentrator (UniScience Univap or Savant Speedvac). To transfer material to the gas-phase sequencer, a small volume of solution should be used to redissolve the peptide; 1% trifluoroacetic acid — acetonitrile (50:50 v/v) is suitable for most peptides. Alternatively, for solid-phase sequencing, SDS-containing buffers compatible with the coupling protocols can be employed (see Chapter 3 of ref. 5).

3. REFERENCES

1. Edman,P. (1950) *Acta Chem. Scand.*, **4**, 283.
2. Morris,H.R. (1980) *Nature,* **286**, 447.
3. Chang,J.Y. (1983) In *Methods in Enzymology.* Hirs,C.H.W. and Timasheff,S.N. (eds), Academic Press, New York, vol. 91, p. 455.
4. Mutamoto,K., Kawauchi,M. and Tuzimura,K. (1978) *Agric. Biol. Chem.*, **42**, 1559.
5. Findlay,J.B.C. and Geisow,M. (eds) (1989) *Protein Sequencing: A Practical Approach.* IRL Press, Oxford.
6. Hewick,R.M., Hunkapiller,M.W., Hood,L.E. and Dreyer,W.I. (1982) *J. Biol. Chem.*, **256**, 7990.
7. Harris,E.L.V. and Angal,S. (eds) (1989) *Protein Purification Methods: A Practical Approach.* IRL Press, Oxford.
8. Brett,M. and Findlay,J.B.C. (1983) *Biochem. J.*, **211**, 661.
9. Aebersold,R.H., Teplow,D.B., Hood,L.E. and Kent,S.B.H. (1986) *J. Biol. Chem.*, **261**, 4229.
10. See Milligen manual for new 6600 solid-phase sequencer.
11. Matsudaira,P. (1987) *J. Biol. Chem.*, **262**, 10035.

CHAPTER 6

Engineering proteins for purification

S.J.BREWER and H.M.SASSENFELD

1. INTRODUCTION

Protein chemists now have the technology to design and synthesize proteins with new properties. Techniques of DNA synthesis and gene splicing are used to engineer cell cultures which produce these proteins. A simple application of this technology is to make proteins which are fused together. These fusion proteins can be given improved stability and purification properties which assist their isolation from culture extracts. Although the purified fusion protein may be used for its immuno- or enzymic activity, it is normally a precursor for a native protein. Therefore, protein chemical techniques are used to hydrolyse a specific peptide bond and release the native protein from the fusion (*Figure 1*).

This chapter discusses the practical considerations needed to design fusion proteins with improved stability and purification properties. However, the design features which allow the native protein to be released from the fusion protein are critical and these are discussed first. Methods are then given which allow the extraction, purification and assay of the fusion protein and the production of the native protein from the fusion protein precursor. Finally, examples are given illustrating these principles.

This technology is the result of the first application of protein engineering. It may be applied in research laboratories to rapidly isolate proteins for structural and functional studies. It may also be used by the biotechnology industry to produce bulk enzyme catalysts and high-purity pharmaceuticals.

2. DESIGN OF FUSION PROTEINS

This section will assist in the design of a fusion protein with improved stability and purification properties by referring to examples from the literature and providing practical guidelines. Genetic engineering techniques can produce proteins fused with simple polypeptides or complete proteins. The genetic manipulations required to produce these engineered proteins are not the subject of this book (for background reading, see ref. 1). However, the protein chemist must work with the genetic engineers to introduce features, such as convenient restriction sites, which allow design changes to be readily made.

The amount of engineered protein produced by cell culture is a product of genetic and physiological factors. The optimization of culture conditions and the use of genetics to modify cellular regulatory systems may allow a single protein to accumulate to 40% of the total cellular protein (for background reading, see ref. 2). However, when even small protein modifications are engineered, large changes in protein expression may

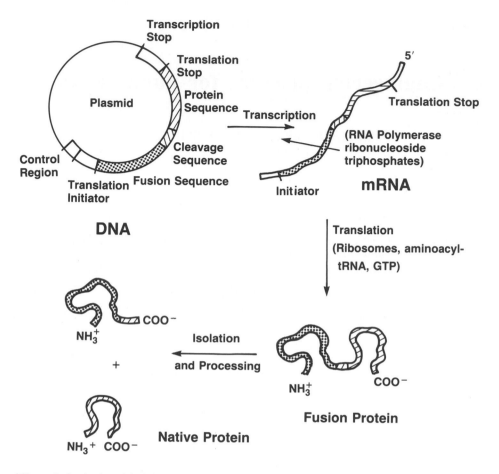

Figure 1. Synthesis and isolation of fusion proteins produced by bacteria.

occur. If a fusion protein is expressed poorly when compared with the native protein, then the problem is unlikely to be intrinsic to the protein itself. Instead, re-optimization of culture conditions may be required. This is achieved by working with genetic engineers and fermentation and cell culture technologists.

2.1 Fusion peptide hydrolysis

A primary consideration in designing a fusion protein is making provision for the ultimate removal of the fusion to produce the native protein. This is required if the fusion protein lacks the desired specificity or biological activity. Hydrolysis can be achieved by incorporating a cleavage peptide which directs chemical or enzymic hydrolysis to the junction of the fusion and protein. Proteases which cleave peptide bonds within polypeptide chains (endopeptidases) or sequentially digest amino acids from the C or N terminus (exopeptidases) and chemical methods can be used (*Table 1*).

In designing a cleavage peptide, the following factors should be considered. First,

Table 1. Cleavage peptides used in fusion proteins.

Peptide	Method	Product
N-Met-C (5)	Cyanogen bromide	N-Met and C
N-Asp-Pro-C (6)	Acid	N-Asp and Pro-C
N-Lys/Arg-C (7)	Trypsin	N-Lys/Arg and C
N-Glu/Asp-C (8)	V-8 protease	N-Glu/Asp and C
N-Lys-Arg-C (9)	Clostropain	N-Lys-Arg and C
N-(Lys)n/(Arg)n (10)	Carboxypeptidase B	n(Lys)/n(Arg) and N
N-Asp-Asp-Lys-C (11)	Enterokinase	N-Asp-Asp-Lys and C
Glu-Ala-Glu-C (12)	Aminopeptidase I	Glu-Ala-Glu and C
N-Ile-Glu-Gly-Arg-C (13)	Factor Xa	N-Ile-Glu-Gly-Arg and C
N-Pro-X-Gly-Pro-C (14)	Collagenase	N-Pro-X and Gly-Pro-C

The amino acid sequences above (where X indicates any amino acid) have been used in fusion proteins to allow a specific hydrolysis reaction. Depending on the hydrolysis method, these cleavage peptides may be used with fusion polypeptides linked at either or both their amino and carboxy termini. These are indicated as N and C polypeptides respectively.

the resistance of the native protein to the various cleavage reactions should be assessed. The efficiency of both chemical and enzymic methods may be affected by the protein's conformation. This can make a protein resistant, even though susceptible amino acids are present. Resistance can be increased by replacing or deleting susceptible amino acids from the native protein. The resulting mutant protein, however, may have altered activity. Secondly, the effects of reaction conditions on the native protein must be considered. Typically, chemical hydrolysis is performed in denaturing conditions which will require the protein to be refolded. Proteases, however, can be used in non-denaturing physiological conditions. Thirdly, the specificity of the reaction should be assessed. Extremes of pH can modify amino acids, and impurities in proteases may cause non-specific hydrolysis. Finally, if large quantities of protein are to be produced by proteolysis, consider if a sufficient quantity of protease with the desired quality is available.

2.2 Fusions for stability

Stabilization of proteins may be achieved by forming an intracellular inclusion body, or exporting the protein into the periplasmic space or culture medium (*Figure 2*). Proteolysis of small polypeptides in the intracellular soluble fraction of *Escherichia coli* is a major factor causing poor yields of extracted protein. Polypeptide fusions can encourage the formation of inclusion bodies in which the protein is denatured and aggregated with other cellular components but stable (*Table 2*). To export proteins, a fusion is made with an export leader sequence which results in its secretion into the periplasm and culture medium. The secretion leader sequence is cleaved by an enzyme in the cell membrane. Examples where proteins have been stabilized by forming fusion proteins are given in *Table 3*.

   To stabilize a protein by making a fusion protein which forms an inclusion body, consider the following factors. An inclusion body may be formed by making a fusion with a portion of a well expressed *E.coli* protein. The resultant protein is unable to fold, and an aggregate, enriched in the fusion protein, forms. However, to maximize

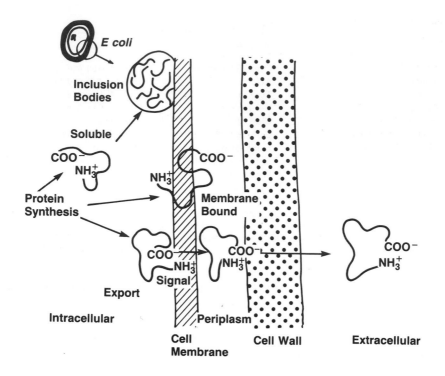

Figure 2. Localization of engineered proteins in *E. coli*.

Table 2. Stability of urogastrone by inclusion bodies formed with basic polypeptide fusions.

Fusion polypeptide	Soluble (%)	Half-life (min)	Insoluble (%)	Stability
None	>90	6	—	—
N-terminal-TrpE	67	25	33	Stable
C-terminal-Polyarg	61	19	39	Stable
TrpE and Polyarg	83	36	17	Stable

The effects of the addition of small, basic polypeptide fusions on the stability of urogastrone when expressed in *E. coli* was studied using pulse-chase experiments (15). The percentage of the total urogastrone expressed, soluble and insoluble, and their corresponding half-lives are shown. The results demonstrate that insoluble proteins in inclusion bodies are stable.

culture productivity and minimize side reactions during release of the native protein, small peptide fusions should be used to stabilize proteins. Also, improving the purification should be considered, by using the isolated inclusion body as an enriched source of fusion protein and enhancing the purification properties of the fusion protein itself (see below). Finally, the protein in the inclusion body will probably be denatured and require refolding. Often this is not a problem with small proteins, but if it is, secreting the protein into the periplasm or culture media using bacterial or yeast expression systems should be considered.

Table 3. Protein fusions for stability.

Polypeptide	Mechanism	Protein
β-Galactosidase (5)	Inclusions	Somatostatin
β-Galactosidase (16)	Inclusions	Insulin A- and B-chains
β-Galactosidase (17)	Inclusions	β-Endorphin
TrpE (15)	Inclusions	Urogastrone
Polyarginine (15)	Inclusions	Urogastrone
Secretion signal (6) and protein A	Extracellular	Insulin-like growth factor (IGF-1)
Secretion signal (18)	Periplasmic	Urogastrone

Table 4. Protein fusions for improved purification.

Purification method	Purification fusion
Ion-exchange	Poly(arg)/(lys)/(lys-arg) (10)
Metal chelate	Histidine-rich polymers (19)
Substrate affinity	β-Galactosidase (14)
	Chloramphenicol aminotransferase
Immunoaffinity	Protein A (6)
	Flag peptides (24)
	Rec A (8)

These examples of purification fusions have been taken from the literature.

2.3 Fusions for purification

Protein purification is achieved by exploiting physicochemical differences in charge, hydrophobicity and size or by using bioaffinity methods based on specific structural or functional properties. By designing fusion proteins to have a unique purification property, a rapid and simple purification can be achieved. In addition, the potential of using the purification fusion for detection or assay of the desired protein should be considered. Spectral, chromatographic, enzymic and antigenic properties may all be used to develop a simple assay. Examples of such purification fusion proteins are given in *Table 4*.

In designing a purification fusion protein, the scale and purpose of the work should first be considered. If a protein is to be isolated on a large scale, then the cost of the purification matrix may be critical. Ion-exchange is particularly suitable for these large-scale operations. If only small quantities are required for experimental evaluation, or the protein has a high value, then more expensive affinity methods are acceptable. Secondly, the presence of residues from the cleavage reaction on the purification of the protein should be considered. Unless the native protein can be readily isolated from the purification cleavage products, many of the advantages of the purification fusion will be lost. Finally, if the protein is produced as an inclusion body, avoid fusions which may interfere with the refolding process. For this reason, a small purification peptide is preferred.

3. EXTRACTION

Engineered proteins may be expressed at different amounts and in different locations within a cell culture. The protein chemist must determine the quantity and location of the desired protein during expression optimization, and be prepared to adapt protein extraction protocols to cope with each possibility. This section will provide a number of methods which can be used to form the starting point for fractionating, extracting and assaying an engineered protein produced by bacterial or yeast cell cultures.

The choice of extraction method will depend on the location of the recombinant protein. Secreted proteins from yeast or bacterial cultures are purified after the cells are removed by either centrifugation or filtration. Proteins which accumulate in the cytoplasm, periplasm, or inclusion bodies require cellular disruption and removal of cell debris before purification. Proteolysis may be a problem with some proteins and low temperatures or protease inhibitors (20) should be used. Centrifugation and disruption give rise to aerosols, and precautions must be taken to avoid inhalation or contamination of the environment.

3.1 Centrifugation of cell cultures

Extracellular proteins will be in the supernatant. Periplasmic, membrane-associated or intracellular proteins will be associated with the cell pellet.

(i) *Analytical*. Dispense 1 ml samples of bacterial or yeast culture into 1.5 ml Eppendorf tubes and cap. Centrifuge at 6000 *g* for 3 min in a bench centrifuge and immediately decant the supernatant.

(ii) *Preparative*. For up to a 10 litre culture, dispense 800 ml of culture into 6 × 1 litre screw cap bottles. Centrifuge at 4°C in a pre-cooled sealed rotor for 30 min at 4000 *g*, for bacterial cultures and 10 min at 4000 *g* for yeast cultures. Decant the supernatant fluid. If the cell fraction is being collected, add remaining culture to the drained bottles and repeat until the culture is harvested.

3.2 Cell disruption

The bacterial or yeast cell pellets are resuspended in a suitable lysis buffer and disrupted, as follows, to release intracellular soluble proteins (see ref. 21 for more details).

(i) *Osmotic shock*. This method is primarily used for the selective release of proteins located in the bacterial periplasm. A sudden change in osmotic pressure is used to disrupt the outer membranous cell envelope. It is useful for analytical scale preparations. Resuspend up to 0.5 ml of packed cells in 4 ml of 20% (w/v) sucrose, 30 mM Tris−HCl pH 8.0, 1 mM EDTA, 0.1% Triton X-100 and leave for 5 min at room temperature. Collect cells by centrifuging at 10 000 *g*, 3 min at room temperature. Resuspend in 2 ml cold water. Swirl gently in ice bath for 10 min, centrifuge at 5000 *g* for 3 min at 4°C and collect the supernatant fluid containing the solubilized protein.

(ii) *Sonication*. Sonication is primarily used for analytical-scale disruption of bacterial cells. Intense ultrasonic waves cause the growth and collapse of vapour bubbles in the cell suspension which lyse bacteria by mechanical shear forces. Efficiency is reduced in highly viscous solutions, or if frothing occurs. A considerable amount of heat is generated which can denature proteins and accelerate proteolytic degradation. Sonication should be in short bursts with intermittent cooling. For analysis, resuspend not more than 0.3 ml of cell paste in 3 ml of ice-cold lysis buffer. Place a 5 mm diameter probe just under the surface and increase the power to the maximum possible without frothing. Sonicate in 3×1 min bursts in an ice bath with 1 min intermittent cooling. For preparative work, suspend 5 ml of cell paste in 50 ml of lysis buffer. With a 10 mm probe, sonicate twice for 4 min at 50 W on ice with cooling for 4 min.

(iii) *Homogenization.* Disruption is accomplished by pumping cells at high pressure through a small orifice against an impaction plate. The pressure drop and stresses cause cell disruption. Efficiency is reduced for high cell densities and about 20% (v/v) packed cells is maximal. This method is particularly useful for pilot-plant-scale extraction of intracellular proteins where large volumes can be pumped from tanks through heat exchangers and into the homogenizer. For the laboratory scale, it is best used in a discontinuous batch mode. Prepare 1 litre of a 20% cell suspension in ice-cold lysis buffer, pump through an APV homogenizer (APV Co.) at 6000 p.s.i., cool and repeat twice.

(iv) *Milling with glass beads.* Yeast cells are typically disrupted by milling with glass beads. For a maximum of 1 ml of yeast cell paste, resuspend in 10 vols of 50 mM Tris−HCl, pH 8 buffer. For each ml of cell suspension add 0.4 ml of glass beads (0.45 mm diameter) and lyse by vortexing 3 times for 30 sec, with intermittent cooling on ice for 30 sec. For preparative work use 20% (v/v) suspension of cells in a ball mill (Dynomill) filled with 0.1 mm glass beads operated for 5 min at 2000 r.p.m. cooled to −20°C with glycol.

3.3 Centrifugation of cell lysate

After centrifugation of the cell lysate, the supernatant will contain soluble intracellular protein and the pellet will contain inclusion bodies and membrane-associated proteins.

(i) *Analytical.* Dispense 1 ml samples of cell lysate into 1.5 ml Eppendorf tubes and cap. Centrifuge at 6000 *g* for 10 min and immediately decant the supernatant.

(ii) *Preparative.* Dispense 200 ml of cell lysate into 250 ml bottles and cap. Centrifuge in a rotor pre-cooled to 4°C at 10 000 *g* for 20 min at 4°C. Carefully decant the supernatant. Wash inclusion bodies or membrane preparations by resuspending the pellets in cold water and repeating the centrifugation step.

3.4 Solubilization of proteins

Engineered proteins which are in the disrupted cell pellet are probably in inclusion bodies (22). Membrane-bound proteins may be solubilized with non-ionic detergents (see Chapter 4). However, extraction of a protein contained in an inclusion body will probably require one of the protein denaturants discussed below. It is common practice to remove soluble cytoplasmic proteins as described above before solubilization of the inclusion body. This approach can result in an extract substantially enriched in the desired protein. Because denaturants are used for solubilization, a protein refolding step will be required either before or after purification (see Section 4).

(i) *Sodium dodecyl sulphate (SDS).* This strong anionic detergent combined with a reducing agent (to break disulphide bonds) and heat, will solubilize nearly all proteins. SDS binds avidly to proteins and complete removal may be difficult. Suspend a cell pellet (30% v/v maximum) in 5% SDS, 40 mM dithiothreitol in 0.1 M Tris−HCl, pH 8.0. Mix thoroughly and heat to 100°C for 15 min.

(ii) *Urea*. High urea concentrations can be used to solubilize proteins. However, urea will form cyanates at alkaline pH values which may derivatize lysine residues in proteins. Use only freshly made ultra-pure urea or de-ionize by filtering through a 100 ml bed of Dowex AG 501-X8 per 10 litre of buffer. Resuspend a cell pellet (20% v/v maximum) in 5 M urea, 40 mM Tris-acetate−NaOH, pH 9.5 and sonicate or homogenize as described above. Centrifuge at 16 000 *g* for 60 min, 20°C. To store, adjust the supernatant to pH 6.0 with 1 M HCl and centrifuge at 16 000 *g* for 60 min at 4°C. At pH values below 6 substantial precipitation of *E.coli* proteins will be observed.

(iii) *Low and high pH*. Extremes of pH have been used to solubilize proteins. Such treatment may cause degradation and modification of proteins and is therefore only recommended when all other methods fail. To a cell pellet, add 10 vols of 0.1 M NaOH. Stir with heating in a boiling water bath and clarify by centrifugation. Adjust to pH 9.0 with acetic acid and centrifuge at 16 000 *g* for 60 min at 4°C.

(iv) *Guanidine*. This is probably the most frequently used solubilizer for inclusion bodies. Guanidine is highly corrosive to stainless steel, thus glass vessels should be used. Resuspend 20% cell paste in 0.6 ml 6 M guanidine−HCl, 0.1 M Tris−HCl, pH 8.0. Sonicate at 25 W for 1 min on ice. For preparative scale, solubilize by heating in stirred glass vessels. Clarify by centrifugation at 13 000 *g* for 10 min, 20°C.

4. REFOLDING

The refolding of proteins is complex and seldom 100% efficient. Optimal conditions are determined empirically. Protein concentration and redox potential are important, as well as ionic strength, pH, time and temperature. Additives such as Tween-20, β-octylglucoside and glycerol have been successfully employed. The conditions below provide starting points for experimentation. For further discussion, see refs 22 and 23.

(i) *Denaturation*. Dissolve protein at 1 mg ml^{-1} in 6 M urea, or 6 M guanidine−HCl, heat in a boiling water bath for 5 min with 25 mM dithiothreitol, 50 mM Tris−HCl, pH 8. Cool and clarify by centrifugation.

(ii) *Renaturation*. Rapidly dilute sample 10- or even 100-fold to 0.1−1.0 mg ml^{-1} into the following renaturation buffers at 37°C, each containing 10% glycerol, 5 mM glutathione (reduced), 0.5 mM glutathione (oxidized). Buffers: (1) 50 mM Tris−HCl, pH 8; (2) 50 mM β-alanine, pH 3.8; (3) phosphate-buffered saline; (4) 50 mM sodium phosphate, pH 6. Store unstoppered for 24 h to allow air oxidation of thiols. Measure biological activity and select optimum buffer. Repeat process to optimize for protein concentration and reducing agent.

5. PURIFICATION OF FUSION PROTEINS

The methodology used will depend on the purification characteristics introduced at the design stage. Details of these methods are described in the companion volume (25).

However, the following considerations should be made to assist in the development of the purification.

5.1 Ion-exchange chromatography

Polycationic fusions may bind nucleic acids, especially if a soluble intracellular protein is to be purified. To avoid reducing the affinity of the fusion protein for an ion-exchanger, pre-treat the sample as described below. Alternatively, HPLC ion-exchangers, such as Mono S (Pharmacia), will usually bind the protein in the presence of excess nucleic acid and enable a single-step purification.

(i) *Precipitation*. Polyethylenimine is a polycation which binds strongly to nucleic acids to form a precipitate. Excessive additions should be avoided to prevent interference with subsequent ion-exchange steps. To a clarified lysate containing 10 mg ml^{-1} protein in 0.1 M Tris−HCl, pH 8, add 0.08% polymin P (w/v) (BDH). Leave at 4°C for 5 min then clarify by centrifugation.

(ii) *Digestion*. At high salt concentrations, nucleic acids will dissociate from proteins, allowing digestion with nucleases. To a clarified lysate containing 2 mg ml^{-1} of protein, add an equal volume of 4 M NaCl and 100 μg DNase I per mg of protein (Sigma). Dialyse for 16 h at 4°C and clarify by centrifugation.

5.2 Affinity chromatography

In some cases the observed affinity will be less than expected. This may occur for several reasons, including low ligand concentration on the column, steric hindrance, or an incomplete binding site on the fusion protein. These problems can be mitigated by careful column preparation and the use of high salt buffers to decrease non-specific interactions. When using immunoaffinity, try to avoid antibodies which require harsh elution conditions.

5.3 Purification of denatured proteins

In some cases the fusion protein will have to be extracted with denaturants such as urea or guanidine. If this occurs, renaturation will almost always be required before undertaking an affinity purification. If the purification fusion does not refold efficiently, then the yields will be reduced. For this reason, small simple fusions, such as polyarginine and the flag peptide (see Section 8), may be advantageous.

In the case of ion-exchange, purification can be undertaken while the protein is denatured in urea. Guanidine extracts can also be diluted with urea to enable binding. After adsorption, protein refolding can be undertaken while the protein is still attached to the ion-exchanger.

6. RECOVERY OF NATIVE PROTEIN

The hydrolysis method used to release the native protein from the fusion protein will have been chosen at the design stage. This section describes a selection of methods for the specific hydrolysis of peptide bonds, along with their limitations. The conditions should be viewed as useful starting conditions for further optimization.

6.1 Chemical cleavage

A limited number of chemical methods can be used to cleave peptide bonds. These reactions usually work best under denaturing conditions. The use of low pH and elevated temperatures can cause substantial deamidation of asparagine and glutamine residues.

(i) *Low pH*. Asp-Pro peptidyl bonds are unusually susceptible to hydrolysis at low pH. In the absence of denaturants, hydrolysis may be unacceptably slow, and steric factors may cause the cleavage at other aspartic acid residues. Depending on whether the fusion is at the carboxy or amino terminus, Asp or Pro will be left on the protein. However, this method has the advantage that the Asp-Pro bond is rare in proteins and selectivity is high. Incubate protein at 0.5 mg ml^{-1} in 7 M guanidine−HCl, 10% acetic acid (v/v) adjusted to pH 2.5 with pyridine, at 56°C for 24 h. Neutralize sample in an ice bath with 1.0 M ammonium hydroxide.

(ii) *Cyanogen bromide (CNBr)*. This reaction hydrolyses the peptide bond on the carboxyl side of methionine. Dissolve protein at 2 mg ml^{-1} in 6 M guanidine− HCl, 0.1 M HCl, pH 1.0 and add an equal weight of cyanogen bromide. Leave for 4 h and neutralize with 5 M ammonium hydroxide.

6.2 Enzymic proteolysis

There are a large number of enzymes able to cleave peptide bonds. These proteases have different optimal conditions for activity and some are even active in denaturants. To achieve the desired specificity, highly purified proteases must be used, or contaminating proteases must be inactivated with selective inhibitors. Immobilized proteases are recommended to improve stability, allow re-use and reduce contamination of the product with protease. Suppliers of these enzymes are Sigma, Boehringer Mannheim, Cooper Biomedical & ICN Immunobiologicals, and Pierce.

6.2.1 *Immobilization of protease*

Two methods are presented for immobilizing proteases on to agarose or silica (for alternative methods see ref. 23). The latter substantially reduces leaching of the protease and is preferred when proteins are intended for clinical use. To maintain immobilized proteins in suspension without damaging the support matrix, slow end-over-end rotation or gentle mixing with an overhead stirrer is recommended.

(i) *Sepharose*. Dissolve 20 mg of protease in 10 ml of 0.1 M sodium bicarbonate buffer, pH 8.3. Add to 10 ml of CNBr-activated Sepharose and react for 16 h at 4°C. Wash with phosphate-buffered saline (PBS) and store in PBS and 0.1% sodium azide at 4°C.

(ii) *Silica*. Suspend 5 g of 50-micron 5000 Angstrom Lichrosphere (Merck) in 1 M HCl, leave overnight, wash on a sintered funnel with water until neutral, then wash with acetone and dry in a fume cupboard. Transfer to a round-bottomed flask fitted with a calcium chloride plug, add 250 ml of 2% aminopropantriethoxy-silane dissolved in toluene and heat for 8 h in a 90°C ultrasonic water-bath with 2 min sonications every 30 min. Wash the silica over a period of 4 h with 150 ml of toluene, 600 ml methanol and 100 ml of diethyl ether and dry. Add 80 ml of 10% glutaraldehyde dissolved in 50 mM sodium phosphate buffer, pH 8.5

and incubate for 4 h at 90°C in the ultrasonic bath as above, then wash in 1 litre of water. Make up the protein coupling buffer of 0.1 M borate, 0.5 M NaCl, 0.01 M MgCl$_2$ and 0.001 M CaCl$_2$ adjusted to pH 8.5 with NaOH and equilibrate the silica in this buffer. Dissolve 100 mg of enzyme in 100 ml of coupling buffer and incubate with the activated silica for 2 h at 25°C in the ultrasonic water-bath as above. Wash with 50 mM sodium phosphate, pH 8.5 buffer, and react at 25°C for 2 h with 200 ml of 5 mM cyanogen bromide then 200 ml of 2 mM sodium borohydride. Store in PBS and 0.1% sodium azide at 4°C.

6.2.2 *Digestion*

Exopeptidases sequentially digest amino acids from either the carboxy or amino terminus. Digestion will stop when a non-reactive amino acid is encountered. In contrast, endopeptidases cleave accessible amino acids within proteins. Careful monitoring of hydrolysis is required to ensure that the correct specificity is achieved. Because acid is released during hydrolysis, the pH needs to be controlled to maintain optimum activity.

(i)   *Carboxypeptidase B.* This exopeptidase sequentially digests arginine or lysine residues present at the C terminus. It is readily available without contaminating proteases, and is very stable after immobilization. Use between pH 5 and 10 at NaCl concentrations of between 50 mM and 2 M. The enzyme is also active in 5 M urea. Immobilize carboxypeptidase on Sepharose as described above. At pH 8.1, use 100 μl of carboxypeptidase-B (CPB)-Sepharose for every mg of protein. Digest in a sealed bottle for 30 min at 22°C, by gentle end-over-end rotation. Remove carboxypeptidase B-Sepharose from the digest by filtration on a sintered glass funnel.

(ii) *Trypsin.* This endopeptidase cleaves on the carboxyl side of accessible arginine and lysine residues. Treatment with tosyl-phenylalanine chloromethylketone, a protease inhibitor, eliminates any contaminating chymotrypsin activity. It is also active in 5 M urea. Cleavage can be limited to arginine by reversibly blocking lysine residues with citraconic anhydride. Dissolve protein at 10 mg ml^{-1} in 100 mM Tris$-$HCl, pH 8.0. Incubate at 37°C with 0.01$-$0.05% (w/w) trypsin. Store trypsin at 10 mg ml^{-1} in 10 mM HCl.

(iii) *V-8 protease.* This enzyme cleaves on the carboxyl side of accessible glutamic or aspartic acid residues. Assay, if required, using CBZ-Phe-Leu-Glu-4-nitroanalide (Boehringer). Dissolve protein in 0.1 M sodium phosphate, pH 7.5, 25 mM EDTA at 4 mg ml^{-1}. Digest with 0.01$-$0.05% (w/w) V-8 protease (25 nM, 0.1$-$1 mM protein) at 37°C for 6 h.

(iv) *Enterokinase.* This protein cleaves on the carboxyl side of lysine in the sequence Asp-Asp-Lys-X and is an example of a protease with an extended active site. Dissolve protein in 10 mM Tris$-$HCl, pH 8 at 2 mg ml^{-1}. Make reaction mixture 40 mM in octyl glucoside and incubate with 0.02$-$0.2 mg ml^{-1} of bovine enterokinase (typically 0.2$-$2% by molarity) for 16 h at 37°C. Store enterokinase in 10 mM Tris$-$HCl, pH 8.0 at -70°C.

6.3 Analysis of digestion

If the protease digestion is not highly specific, the digestion reaction must be monitored.

This is achieved by correlating the release of acid with the release of product on a pilot scale, followed by a full-scale digestion. The protein sample must be in a low-molarity buffer solution (less than 5 mM) and all solutions must be at controlled temperatures. A pH-stat is used to monitor the kinetics of acid production, whilst incubating with the protease. At suitable times, samples are removed and analysed for the appearance of product using a suitable HPLC assay. The optimum digestion time for maximum product is determined, noting the amount of base consumed at this time. The digestion is repeated at the desired scale, stopping the reaction when the calculated equivalents of base have been added.

7. ASSAY

Many purification fusions can provide a sensitive and quantitative assay for the fusion protein. When affinity fusions such as chloramphenicol aminotransferase and β-galactosidase are used, this is accomplished by an enzymic assay with the appropriate substrate. For ion-exchange fusions such as polyarginine, HPLC can be used, or the arginine released after carboxypeptidase-B digestion can be measured directly. Immunogenic fusions can be assayed by standard immunological methods. When removal of the fusion is required, such assays can be used to determine the efficiency of cleavage.

7.1 **HPLC assay**

HPLC assays will depend on the exact nature of the fusion protein. The following protocol is suitable for charged purification fusions. For ion-exchange, urea-solubilized proteins can be analysed directly.

(i)   Dilute 0.1 ml of extract with 0.9 ml of an appropriate low-salt buffer and clarify by centrifugation or filtration.

(ii)   For basic proteins use a 5 × 50 mm SP-TSK-5PW (Bio-Rad) HPLC column at 20°C. Equilibrate in 20 mM Mes-NaOH buffer, pH 6 and elute with a 30 ml linear gradient (0−1.5 M NaCl) at 1 ml min^{-1}.

(iii)   For acidic proteins use a DEAE-5PW column in 20 mM Tris−HCl, pH 8.0.

(iv)   For urea solubilized proteins, make each buffer up in 6 M urea, ultra-pure or deionized, and use within 48 h.

7.2 **Immunoassay**

A typical ELISA assay suitable for immunogenic peptides is described.

(i)   Apply sample and standard solutions to haemagglutination (HA) plates (Millipore) at a concentration of 40 ng per well and incubate for 30 min at room temperature.

(ii)   Block non-specific protein binding sites by incubating with Tris-buffered saline, pH 7, in 3% bovine serum albumin (BSA) for 1 h at room temperature.

(iii)   Add a pre-determined quantity of mouse monoclonal antibody and incubate the plates for 1 h.

(iv)   Wash with PBS and add an alkaline phosphatase-labelled goat anti-mouse antibody (Sigma).

(v)   Following a 1 h incubation, wash the plates several times with PBS and then add the colorimetric reagent (substrate tablets, Sigma).

(vi)   Transfer contents of each HA plate to a polystyrene 96-well plate and measure the absorbance at 405 nm on a Titerscan (Flow Laboratories).

8. EXAMPLES

This section describes the use of fusion proteins to assist purification. In the first example both C- and N-terminal polypeptide fusions are used to stabilize and improve the purification properties of a small polypeptide hormone, urogastrone, produced as inclusions by *E.coli*. A purification scheme is described which yields highly purified authentic human urogastrone suitable for clinical studies. The second example uses a C-terminal Arg-Lys fusion to assist the purification of a soluble intracellular *E.coli* enzyme using ion-exchange chromatography. This technique is suitable for the production of large quantities of highly purified proteins. The final example describes the use of a small peptide fusion which allows the purification of a colony-stimulating factor (CSF-1) from the extracellular media of yeast cultures by immunoaffinity chromatography. This method enables the production of high-purity proteins for research.

8.1 **Purification of β-urogastrone**

Urogastrone (human epidermal growth factor) is a 53-amino-acid polypeptide hormone with ulcer- and wound-healing properties. The protein can only be isolated in minute quantities from human urine. *E.coli* has been engineered to produce this protein, but it is rapidly degraded. Its stability and purification properties, however, are improved when a modified gene is used to produce urogastrone with a N-terminal fusion from the TrpE protein and a C-terminal fusion of six arginine residues.

A purification of β-urogastrone is illustrated in *Table 5*. The urogastrone fusion protein accumulates in inclusion bodies, where it is resistant to proteolysis (*Table 2*). The fused protein requires solubilization in urea when the C-terminal polyarginine fusion allows a highly purified intermediate to be isolated by ion-exchange chromatography. Digestion with trypsin releases the native β-urogastrone from the fusion protein precursor (*Figure 3*). Although the protein is resistant to trypsin, over-digestion can occur, and careful monitoring is required. Final purification is achieved using preparative HPLC to yield gram quantities of highly purified β-urogastrone.

Table 5. Typical purification of urogastrone from a fusion protein precursor.

Sample	Volume (ml)	Protein (mg)	Urogastrone (mg)	
			Fused	Beta
Extract	2100	15 500	890	–
Refold	1700	8600	760	–
Column 1	116	820	730	–
Trypsin digest	130	820	–	360
Column 2	56	268	–	270

The purification method is described in Section 8.1.

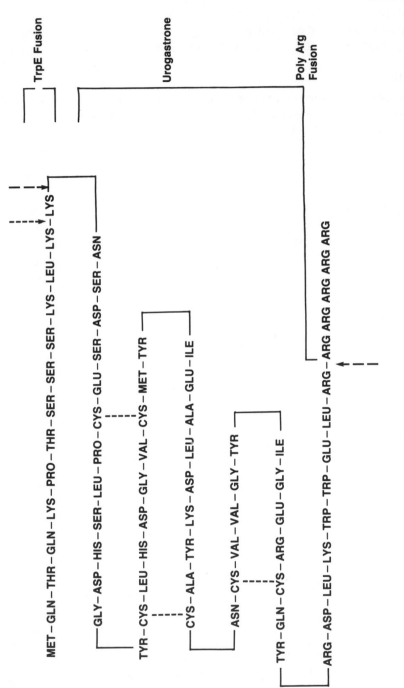

Figure 3. Trypsin cleavage of β-urogastrone fusion protein. The desired trypsin cleavage sites to obtain β-urogastrone from a fusion protein precursor are indicated (– – –→).

8.1.1 *Extraction*

(i) *Growth of cultures*. Culture *E.coli* cells in 8-litre fermenters containing modified M9 media (3) to an absorbance of approximately 3.0 at 670 nm.

(ii) *Disruption of cells*. Harvest cells by centrifugation, disrupt by homogenization in 1 litre of 0.1 M Tris−HCl, pH 8.0, and recover inclusion bodies by centrifugation as described in Sections 3.1−3.3

(iii) *Solubilization*. Suspend inclusion bodies in 1 litre of 5 M urea, 40 mM Tris-acetate−NaOH, pH 9 lysis buffer, sonicate and clarify by centrifugation as described in Section 3.3.

8.1.2 *Refolding and oxidation*

Urogastrone refolds and oxidizes in the presence of 5 M urea.

(i) Leave solubilized inclusion bodies in lysis buffer at 4°C for 24 h.
(ii) Adjust to pH 5.5 with HCl and clarify by centrifugation.

8.1.3 *Pre-treatment and chromatography*

(i) *Pre-treatment*. Precipitate nucleic acids with polymin P as described in Section 5.1.

(ii) *Chromatography*. Apply to a 100 ml column of SP-Sepharose fast-flow equilibrated in 40 mM Tris-acetate, 5 M urea, pH 5.5, wash with this buffer and elute urogastrone with 400 mM NaCl in equilibration buffer.

(iii) Concentrate to 8 mg ml^{-1} by ultrafiltration and dialyse against 40 mM Tris-acetate buffer, pH 5.5.

8.1.4 *Enzyme digestion and rechromatography*

(i) *Enzyme digestion*. Perform pilot digestions with 10 mg of urogastrone with 1:1000 trypsin to protein and analyse for urogastrone production using cation-exchange HPLC (*Figure 4*) followed by a full-scale digestion as described in Sections 6.2 and 6.3.

(ii) *Rechromatography*. Purify β-urogastrone by preparative HPLC (4).

8.2 Purification of aminotransferase

C-terminal poly-Arg-Lys fusions combined with carboxypeptidase B (CPB) digestion allow native proteins to be isolated in a highly purified form by cation-exchange chromatography (*Figure 5*). This fusion has been used to purify bacterial aspartate aminotransferase (EC 2.6.1.1).

Aminotransferase has been cloned and expressed in *E.coli* fused to an additional seven positively charged amino acid residues at the C terminus. A representative purification of aspartate aminotransferase is shown in *Table 6*. The advantage with CPB digestion is that it specifically digests C-terminal basic amino acids (Arg or Lys), then stops. Therefore repeated CPB digestions of the fusion protein do not cause a significant change in enzyme activity. The fusion protein also has the same specific activity as aspartate aminotransferase. Possibly, this is because the C-terminal fusion does not interfere with protein folding, which begins at the N terminus.

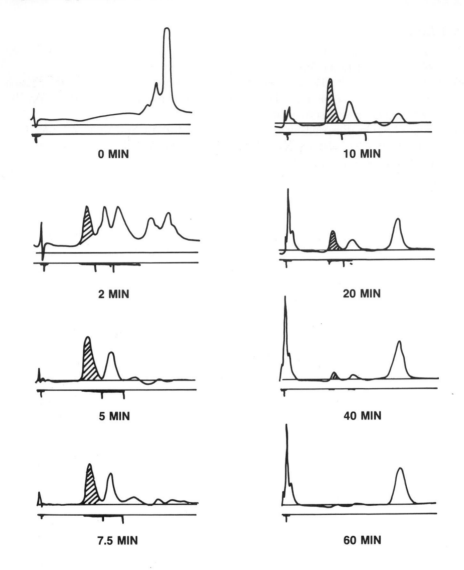

Figure 4. Trypsin digestion of urogastrone fusion protein. Ion-exchange HPLC was used to analyse the digestion of fused β-urogastrone protein. β-Urogastrone product (shaded area) yield was optimal after 5−7 min of digestion. Further incubation resulted in degradation of the protein.

The effectiveness of this purification scheme is illustrated by the SDS−PAGE gel shown in *Figure 6*. Most of the contaminants pass through the column without binding because the pH value chosen is above the pI of most bacterial proteins. However, the fusion protein is retained because of its unusual positive charge. All bound protein is then eluted in a single step and treated directly with CPB. Following dialysis, the aminotransferase passes through the same column unbound and essentially pure. When this column is eluted for a second time, all the contaminants are seen, demonstrating

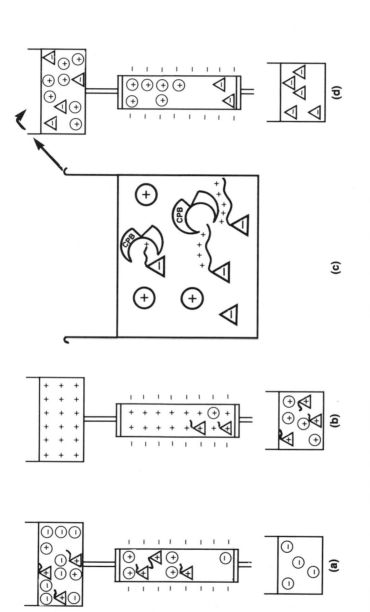

Figure 5. Principle of the polyarginine purification fusion. (a) When an extract containing a mixture of basic (⊕), acidic (⊖) and polyarginine fused (△) proteins are applied to a cation-exchange column, only basic proteins bind. (b) The basic protein contaminants and polyarginine fused proteins can then be eluted with salt or acid. (c) Carboxypeptidase B (CPB) digests the polyarginine purification fusion to produce the native protein, which becomes acidic (△). (d) The basic protein contaminants still bind to the cation-exchange column while the native protein passes through the column and is purified.

Table 6. Purification of aspartate aminotransferase with a basic C-terminal purification fusion.

Sample	Volume (ml)	Protein (mg)	Enzyme (unit)	Activity (unit mg^{-1})
Extract	400	2460	70 100	28
Column 1	300	662	54 300	82
Digest	270	575	46 000	80
Column 2	500	275	39 800	145

The purification method is described in Section 8.2.

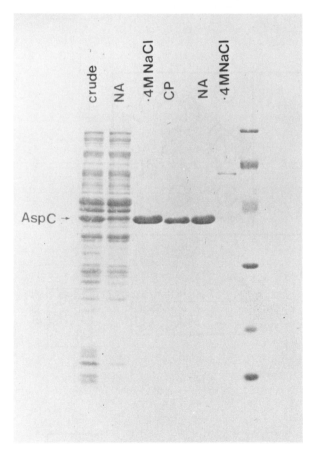

Figure 6. SDS−PAGE illustrating the purification of aminotransferase. Tracks labelled from left to right contained the following samples. Crude, applied crude lysate; NA, unbound protein, first ion-exchange column; 0.4 M NaCl, eluate first ion-exchange column; CPB, carboxypeptidase B digested eluate; NA, digested, unbound aminotransferase from second ion-exchange column; 0.4 M NaCl, eluate second ion-exchange column. Last track, molecular weight standards. AspC indicates position of the aminotransferase band.

that only the fusion protein is affected by CPB digestion. When a control preparation containing undigested aminotransferase is reapplied to the ion-exchanger, it binds and elutes with the contaminants as before.

8.2.1 *Extraction*

(i)   *Growth of cultures*. Inoculate shake flasks (250 ml) containing M9 salts and
        casamino acids (3) directly with 10 μl of recombinant *E. coli* from a glycerol
        stock. Grow cells overnight in an orbital incubator and use to inoculate a 10
        litre fermenter. Grow in the same medium to an absorbance at 670 nm of 2.5.
(ii)   *Disruption of cells*. Harvest culture by centrifugation, disrupt by homogenization
        in 500 ml of 50 mM Mes-NaOH lysis buffer and clarify by centrifugation as
        described in Sections 3.1 – 3.3. The supernatant contains the soluble intracellular
        enzyme for ion-exchange chromatography.

8.2.2 *Ion-exchange chromatography*

(i)   Apply clarified lysate (2 ml) at 20°C to a 50 × 200 mm SP-Sephadex cation-
        exchange column at 10 ml min^{-1}.
(ii)   Wash column with 1 litre of 50 mM Mes-NaOH, pH 6.5 (equilibration buffer)
        and elute with 1 litre of this buffer containing 0.75 M NaCl.

8.2.3 *Digestion of polyarginine and rechromatography*

(i)   Digest partially purified aspartate aminotransferase with 50 ml of CPB-Sepharose
        for 45 min at 20°C as described in Section 6.2.2.
(ii)   Dialyse for 2.5 h against 10 litres of equilibration buffer and rechromatograph
        as described in Section 8.2.2. The non-adsorbed fraction contains purified
        aminotransferase.

8.3 Purification of flag-peptide colony-stimulating factor 1

Flag peptide is designed for immunoaffinity purification. It has the sequence
AspTyrLysAspAspAspAspLys, which includes the enterokinase cleavage site (*Table
1*). A fused protein containing the yeast secretion sequence, flag peptide and colony-
stimulating factor 1 (CSF-1) is secreted by yeast cells. The secretion sequence is cleaved
on export by a protease in the cell membrane and the flag-CSF-1 fusion protein
accumulates in the culture medium.

Immunoaffinity chromatography allows the purification of flag-CSF-1 from a yeast
culture supernatant with a 300-fold increase in specific activity (*Table 7*). The native
protein can be produced by cleavage with enterokinase, but the fusion protein itself
shows full biological activity.

Table 7. Purification of flag-colony-stimulating factor by immunoaffinity chromatography.

Sample	Volume (ml)	Protein (mg)	Units ($\times 10^5$)	Activity (unit mg^{-1})
Supernatant	1000	150	5.5	3.7×10^3
Non-adsorbed	1050	148	1.8	–
pH 7 wash	50	0.2	1.3	–
pH 3 elution (peak)	1.5	0.6	7.0	1.2×10^6

The purification method is described in Section 8.3.

As a result of the highly charged nature of the flag peptide, the antibody binding is calcium-dependent. This has the advantage of allowing elution from the immunoaffinity column to be achieved by removing calcium. Thus, proteins which are sensitive to pH may be eluted at neutral pH in buffers free of calcium. The hydrophilic nature of the flag peptide also ensures that it is accessible to immobilized antibodies for purification and is available to cleavage by enterokinase.

8.3.1 *Extraction*

(i)   *Growth of cultures*. Culture recombinant *Saccharomyces cerevisiae* on complex media in 10-litre fermentors at 30°C until stationary phase of growth.

(ii)   *Clarification*. Centrifuge as described in Section 3.1 retaining the culture supernatant. Filter the supernatant through a 0.45-μm cellulose acetate filter.

8.3.2 *Immunoaffinity chromatography*

(i)   To the filtered supernatant (1 litre) add Hepes − NaOH, pH 7.0, $CaCl_2$ and NaCl to 50 mM, 0.1 mM and 0.15 M respectively and re-adjust to pH 7.0.

(ii)   Apply to a 1 ml immunoaffinity column at 1 ml min^{-1} overnight.

(iii) Wash the column with 50 ml of PBS buffer containing 0.1 mg ml^{-1} $CaCl_2$ then elute flag-CSF-1 with 10 ml of 0.1 M glycine − HCl buffer, pH 3.0.

8.3.3 *Cleavage of flag peptide*

(i)   Make the pooled flag-CSF-1 10 mM in Tris − HCl, pH 8.0, and adjust to pH 8.0 with 1 M NaOH.

(ii)   Digest with enterokinase as described in Section 6.2.2.

9. REFERENCES

1. Old,R.W. and Primrose,S.B. (1985) *Principles of Genetic Manipulation, An Introduction to Genetic Engineering*. University of California Press, Berkeley.
2. Carrier,M.J., Nugent,M.E., Tacon,W.C.A. and Primrose,S.B. (1983) *Trends in Biotechnology*, **1**, 109.
3. Sassenfeld,H.M. and Brewer,S.J. (1984) *Bio/Technology*, **2**, 76.
4. Brewer,S.J., Dickerson,C.H., Ewbank,J. and Fallon,A. (1986) *J. Chromatogr.*, **362**, 443.
5. Itakura,K., Hirose,T., Crea,R., Riggs,A.D., Heyneker,H.L., Bolivar,F. and Boyer,H.W. (1977) *Science*, **198**, 1056.
6. Nilsson,B., Holmgren,E., Josephson,S., Gatenbeck,S., Philipson,L. and Uhlen,M. (1985) *Nucleic Acids Res.*, **13**, 1151.
7. Smith,J., Cook,E., Fotheringham,I., Pheby,S., Derbyshire,R., Eaton,M.A.W., Doel,M., Lilley,D.M.J., Pardon,J.F., Patel,T., Lewis,H. and Bell,L.D. (1982) *Nucleic Acids Res.*, **10**, 4467.
8. Bittner,M.M., Goldberg,S., Heeren,B. and Galluppi,G. (1986) *Fed. Proc.*, **45**, 3049.
9. Bennett,A., Rhind,S.K., Lowe,P.A. and Hentschel,C. (1984) International Patent WO 84 04756.
10. Brewer,S.J. and Sassenfeld,H.M. (1985) *Trends in Biotechnology*, **3**, 119.
11. Mayne,N.G., Burnett,J.P., Belegaje,R. and Hsiung,H.M. (1983) European Patent EP 85361.
12. Dalbøge,H., Dahl,H.H.M., Pederson,J., Hansen,J.W. and Christenson,T. (1987) *Bio/Technology*, **5**, 161.
13. Nagai,K. and Thøgersen,H.C. (1984) *Nature*, **309**, 810.
14. Germino,J. and Bastia,D. (1984) *Proc. Natl. Acad. Sci. USA*, **81**, 4692.
15. Sassenfeld,H.M. (1986) Ph.D. Thesis, University of Liverpool, UK.
16. Goeddel,D.V., Kleid,D.G., Bolivar,F., Heyneker,H.L., Yansura,D.G., Crea,R., Hirose,T., Kraszewski,A., Itakura,K. and Riggs,A.D. (1979) *Proc. Natl. Acad. Sci. USA*, **76**, 106.
17. Shine,J., Fettes,I., Lan,N.C.Y., Roberts,J.L. and Baxter,J.D. (1980) *Nature*, **285**, 456.

18. Oka,T., Sakamoto,S., Miyoshi,K., Fuwa,T., Yoda,K., Yamasaki,M., Tamura,G. and Miyake,T. (1985) *Proc. Natl. Acad. Sci. USA,* **82**, 7212.
19. Smith,M.C., Furman,T.C., Ingolia,T.D. and Pidgeon,C. (1988) *J. Biol. Chem.,* **263**, 7211.
20. Beynon,R.J. (1989) In *Protein Purification Methods: A Practical Approach.* Harris,E.L.V. and Angal,S. (eds), IRL Press, Oxford, p. 40.
21. Salusbury,T. (1989) In *Protein Purification Methods: A Practical Approach.* Harris,E.L.V. and Angal,S. (eds), IRL Press, Oxford, p. 87.
22. Marston,F.A.O. (1986) *Biochem. J.,* **240**, 1.
23. Angal,S. and Dean,P.D.G. (1989) In *Protein Purification Methods: A Practical Approach.* Harris, E.L.V. and Angal,S. (eds), IRL Press, Oxford, p. 245.
24. Hopp,T.P., Prickett,K.S., Price,V.L., Libby,R.T., March,C.J., Cerretti,D.P., Urdal,D.L. and Conlon,P.J. (1988) *Bio/Technology,* **6**, 1204.
25. Harris,E.L.V. and Angal,S. (eds) (1989) *Protein Purification Methods: A Practical Approach,* IRL Press, Oxford.

CHAPTER 7

Example purifications

1. INTRODUCTION

This chapter describes a number of protein purifications which serve as examples of the tremendous variety of methods applicable to purify different proteins. Basic principles underlying each technique have been covered in the companion volume (1) and specific applications are covered by previous chapters. The examples in this chapter were selected to illustrate aspects of designing a purification scheme, while at the same time being of direct use to many researchers in fields of growing popularity.

Section 2 is on cytochromes, a class of proteins with functional similarities, but differing enormously in their physical properties, such that they are considered under three separate sub-sections. The author has provided a step-by-step guide to effective planning which is particularly useful for the novice attempting a protein purification.

Section 3, on growth factors, describes the problems associated with the purification of this class of proteins, which are typically found at extremely low abundance. The problems have been largely overcome by the production of recombinant growth factors.

Section 4 outlines methods for the purification of collagen types. The proteins are unusual in that they are generally insoluble and must be solubilized by controlled proteolysis before purification can begin.

Section 5 describes methods for the simultaneous isolation of three quite similar enzymes. Unlike some of the more widely known examples of isoenzymes, these metalloproteinases are present at very low abundance and appear to autoactivate and autodegrade very easily.

The final section includes detailed methods for the purification of monoclonal antibodies, which are now commonly available, as well as methods for preparation and purification of antibody fragments, which would be of general use but rarely covered comprehensively in immunology texts.

2. PURIFICATION OF CYTOCHROMES—S.J.Froud

Cytochromes are a ubiquitous and diverse group of proteins. They range from the small (8000 daltons), soluble, single-subunit cytochromes *c* of bacteria to the large (0.5 million daltons) transmembrane respiratory complexes of mammals. It is not within the scope of this section to survey all of the methods employed to purify such a diverse group of proteins. Three protocols will be discussed. The procedure for the purification of the mammalian cytochrome oxidase represents the traditional methods of fractionation by precipitation. In contrast, the protocols presented for the purification of the bacterial cytochromes draw most heavily upon chromatographic separations on the basis of size and charge. The use of ionic and non-ionic detergents in the solubilization of the two terminal oxidases, mammalian cytochrome oxidase and cytochrome *co* from *Methylophilus methylotrophus*, is also discussed.

In order to obtain an estimate of the time required to devise a protocol for the purification of a protein, it is necessary to perform a preliminary characterization such as that described in *Method Table 1*. In general, a protein that is easily assayed, abundant, soluble and stable can be purified in a few months. Several months should be added to the time estimate for each negative answer to the questions in *Method Table 1*.

Method Table 1. Preliminary characterization.

All samples should be maintained at 4°C.

A.   *Is a rapid assay available?*

B.   *Is the protein abundant in the source material?*

C.   *Is the protein soluble?*
  1.   Measure the amount of target protein in source material (if possible).
  2.   Disrupt tissue or bacterium. Centrifuge at 6000 g for 10 min. Resuspend pellet (cell debris) in an equal volume of buffer.
  3.   Divide the supernatant (crude cell extract) into three aliquots and determine the location of the target protein by assaying all of the cell extracts (cell debris, crude cell extract, soluble cell extract, membrane pellet, KCl treated soluble extract and pellet):
     (i)   Cell debris: if the majority of the protein is found in this extract then either the protein is in an insoluble matrix or the disruption regime is ineffective.
     (ii)   Centrifuge crude cell extract at 135 000 g (30 min). Recover supernatant (soluble cell extract, containing the soluble proteins) and resuspend pellet (containing membrane-associated proteins) in an equal volume of identical buffer.
     (iii) Dissolve solid KCl slowly into crude cell extract to give a final concentration of 0.5 M KCl. Centrifuge and recover treated soluble cell extract (containing peripheral membrane proteins) and treated membrane fraction (containing integral membrane proteins) as in (ii). Treatment with 0.5 M KCl will dissociate peripheral membrane proteins. These proteins, which are loosely associated with the membrane surface, may be treated as soluble proteins once dissociated. The catalytic activity of enzymes, however, may be adversely affected by the presence of 0.5 M KCl. If the activity of this fraction is unexpectedly low, remove the KCl by dialysis prior to assay.

D.   *Is the protein stable in solutions of high ionic strength?*
  1.   Prepare a series of samples in salt solutions: for example 0.2 M and 0.5 M sodium chloride, magnesium chloride, ammonium acetate and ammonium sulphate.
  2.   Divide samples into two aliquots. Dialyse the first to remove the salt.
  3.   Assay.

E.   *Is the protein stable upon storage?*
1.   If they contain the target protein, divide all of the samples produced in procedures *C* and *D*, above, into two aliquots.
2.   Store one aliquot at 4°C and the other at ambient temperature for 24 h.
3.   Re-assay.

F.   *Is the protein stable to changes in pH?*
1.   Prepare cell extract that contains the target protein, as described in (*C*), above.
2.   Incubate samples in a series of buffers. Leave one sample untreated.
3.   Assay all samples at the same pH in a strong assay buffer after 1, 4 and 24 h.

2.1 Purification of the soluble cytochromes *c* from *Methylophilus methylotrophus*

2.1.1 *Preliminary questions*

These cytochromes have all the advantages required for the rapid development of a purification protocol; all of the questions in *Method Table 1* can be answered in the affirmative. The basic protocol for the purification of these soluble, stable and easily assayed proteins was developed in a few months (2). The established purification procedure can be completed routinely in two weeks.

One of the requirements for the swift development of a purification protocol is the availability of a rapid assay. The cytochromes *c* are routinely detected by their absorption at 550 nm (see *Method Table 2*). Unless the solutions are turbid, it is not necessary to record difference spectra. Samples can be screened at the rate of 30 per hour. These cytochromes have no catalytic activity.

The cytochrome contents of bacteria vary greatly in response to the growth environment. Selection of a suitable growth regime is an important step in a programme to purify a bacterial protein. Analysis of samples taken throughout batch culture under a number of potentially suitable growth conditions will allow the optimum time for harvesting to be determined. In this case the organism produces maximal amounts of the cytochromes *c* when grown on methanol in continuous culture under conditions of oxygen limitation (3). Under these growth conditions *Methylophilus methylotrophus* produces two soluble cytochromes *c*, designated c_H and c_L. Both of these are purified by the procedure described below. A third cytochrome, designated c_{LM}, is produced during carbon-limited growth. It may be purified by modifying these procedures (2).

2.1.2 *Protocol*

The procedures are described in *Method Tables 3* and *4*.

The stability of these cytochromes allows the use of standard harvesting and storage procedures. Similarly, sonic disruption, a relatively harsh procedure, may be used if convenient.

The procedure as originally developed (2), involved precipitation by acidification to pH 4.2. If the pH falls to 4.0 during the acid precipitation step, however, there is an unacceptable decrease in the yield. To minimize losses during routine purifications,

Method Table 2. Assay of cytochromes by absorption spectroscopy.

Record the spectra in a split beam spectrophotometer. Routine assays will be made more rapid if a suitable single wavelength (or wavelength pair) can be determined. Examples of absorption spectra are presented in *Figures 1* and *2*.

A. *(Reduced) minus (oxidized) difference spectra*

1. Divide a 2 ml sample into two cuvettes.
2. Add 1−4 crystals of sodium dithionite to reduce the first sample.
3. Add 1−5 μl of 3% v/v hydrogen peroxide or 1−2 crystals potassium ferricyanide to oxidize the reference sample.
4. Mix and record spectrum from 250 nm to 650 nm if using hydrogen peroxide. Mix and record spectrum from 500 nm to 650 nm if using potassium ferricyanide.

Characteristic α-absorption peaks:

Cytochrome *a*,	600−605 nm.	Cytochrome *c*,	547−556 nm.
Cytochrome a_1,	585−598 nm.	Cytochrome *d*,	625−635 nm.
Cytochrome *b*,	555−565 nm.		

Cytochrome a_2 has been renamed cytochrome *d*. Cytochrome *f* and *o* are *b*-type cytochromes. Spectra for cytochromes also exhibit a strong (soret) absorption peak between 400−450 nm and usually a weak β-absorption peak between 500−530 nm. For examples see *Figures 1* and *2*.

B. *(Reduced plus carbon monoxide) minus (reduced) difference spectra*

1. Add 1−4 crystals of sodium dithionite to reduce a 2 ml sample.
2. Divide sample into two cuvettes.
3. Under conditions of reduced lighting, gently bubble carbon monoxide (CO) through the first sample for 30 s.
4. Simultaneously bubble nitrogen through the reference sample.
5. Record spectra immediately and, for the detection of cytochromes that bind CO slowly, at intervals for up to 60 min.

Spectral characteristics:

Cytochrome a_3:	Trough at approximately 440 nm.
Cytochrome *o*:	Troughs at approximately 560 nm and 430 nm.
Cytochrome c_{CO}:	Troughs at approximately 550 and 420 nm.

Cytochromes c_{CO} usually bind CO slowly.

For examples see *Figure 2*.

C. *Typical extinction coefficients for cytochromes*

The method for recording (dithionite reduced) minus (ferricyanide oxidised) difference spectra are described in (**A**), above. The extinction coefficient is calculated from the

Beer-Lambert Law:

$$E = \frac{A_S - A_R}{c \cdot L}$$

where c is the light path (cm), and L = concentration (M), calculated from the protein concentration and molecular weight

Cytochrome	Extinction coefficient (E)	Absorption wavelength (A_S)	Reference wavelength (A_R)
a	13 500 − 16 000	605 nm	630 nm
b	22 000	556 nm	575 nm
c	17 000 − 21 000	550 nm	650 nm
o (CO-binding)	170 000	417 nm	432 nm

Method Table 3. Preparation of soluble cell extract and membrane fraction from *Methylophilus methylotrophus*.

Operations are performed at 4°C.

1. (Day 1). Harvest bacteria by centrifugation (10 000 g, 20 min).
2. Wash bacteria by resuspending pellet to original volume using ice-cold 25 mM Mops − NaOH buffer (pH 7.0) and re-centrifuging.
3. Use immediately or freeze in thin pancakes by immersion in liquid nitrogen. Store at −20°C.
4. Resuspend 200 g of thawed bacteria (wet weight) in the same buffer to a final volume of 300 ml. Add 50 mg of deoxyribonuclease.
5. Disrupt bacteria by passing twice through a French pressure cell at 100 MPa. Alternatively, a 100 W ultrasonic disintegrator at 20 kHz may be used (4 × 1 min cycles).
6. Dilute with the same buffer to 600 ml. Centrifuge (6000 g, 10 min).
7. Centrifuge supernatant (crude cell extract) at 135 000 g (90 min).
8. Carefully remove supernatant (soluble cell extract). Use immediately or store at −20°C. Gently resuspend pellet (membrane fraction) after adding an equal volume of the same buffer. Use immediately or store in liquid nitrogen.

therefore, the pH is adjusted to pH 5.0 only. This step can be omitted altogether, although this may result in a requirement for repetition of the final stages (with slight modifications) to obtain the desired purity (4).

The initial DEAE-cellulose column serves to resolve the two cytochromes and allows the same assay to be used for each in the subsequent stages. At this stage a relatively large column is required compared to that needed for subsequent ion-exchange steps

Method Table 4. Purification of the soluble cytochromes *c* of *Methylophilus methylotrophus*.

Operations are performed at 4°C. The eluate from chromatography columns is collected as 10 ml fractions.

A. *Precipitation by acidification*

1. (Day 2). Acidify 500 ml of soluble cell extract (see *Method Table 3*) by dropwise addition of 1 M HCl while stirring. Bring to pH 5.0 and then stir for a further 5 min. Centrifuge at 10 000 *g* (10 min).
2. Adjust pH of supernatant to pH 8.0 by dropwise addition of 1 M NaOH while stirring.

B. *DEAE-cellulose chromatography*

1. Apply acid-treated extract to a DEAE-cellulose (Whatman DE-52) column (5 × 15 cm) equilibrated with 20 mM Tris−HCl buffer (pH 8.0).
2. Recover the cytochrome *c* not adsorbed (cytochrome c_H) and store at −20°C.
3. Elute the adsorbed cytochrome *c* (cytochrome c_L) overnight using a linear gradient up to 200 mM Tris−HCl buffer (pH 8.0) containing 200 mM NaCl, with a total elution volume of 1 litre. The cytochrome c_L will elute at the approximate mid-point of the gradient.

C. *Cytochrome c_L: size-exclusion and DEAE-cellulose chromatography*

1. (Day 3). Concentrate the eluate to 20 ml while under nitrogen using a 5000 mol. wt cut-off ultra-filtration membrane (Ulvac G-05-t).
2. Apply to an upward-flow Sephadex G-150 column (3.5 × 80 cm) equilibrated with 100 mM Tris−HCl (pH 8.0).
3. (Day 5). Pool the fractions containing the cytochrome *c*. Reduce the ionic strength by stirring gently while diluting with an equal volume of distilled water.
4. Apply to a DEAE-cellulose column (3 × 9 cm) equilibrated with 50 mM Tris−HCl buffer (pH 8.0).
5. Elute overnight with a linear gradient up to 125 mM Tris−HCl (pH 8.0) containing 125 mM NaCl, with a total elution volume of 1 litre. The cytochrome c_L will elute near the end of the gradient.
6. (Day 6). Concentrate to 15 ml as described in C1, and apply to an upward flow Sephadex G-75 Superfine column (2.4 × 85 cm) equilibrated with 100 mM Tris−HCl (pH 8.0).
7. (Day 8). Pool fractions containing the purified cytochrome *c*. Concentrate if required (see C1, above) and/or change buffer (by dialysis or size-exclusion chromatography), and store at −20°C.

D. *Cytochrome c_H: size-exclusion and CM-cellulose chromatography*

1. (Day 4). Pool the fractions containing cytochrome c_H from step *B*2, above. Concentrate and perform size-exclusion chromatography as described in steps *C*1 and *C*2, above.

2. (Day 6). Pool the fractions containing the cytochrome. Dialyse against 2 litres of 5 mM sodium acetate buffer (pH 5.6) using an ultrafiltration cartridge with a molecular weight cut-off at 200 daltons.

3. Apply to a CM-cellulose (Whatman CM-52) column (3 × 10 cm) equilibrated with the acetate buffer.

4. Elute the cytochrome with a linear gradient up to 120 mM sodium acetate buffer (pH 5.6), with a total elution volume of 1 litre. The cytochrome c will elute near the mid-point of the gradient.

5. (Day 7). Pool the fractions containing the cytochrome c, and concentrate to 15 ml as described in $C1$, above. Apply to an upward flow Bio-Gel P10 size-exclusion column (2.4 × 80 cm) in 100 mM Tris−HCl (pH 8.0).

6. (Day 8). Pool fractions containing the purified cytochrome c_H. Concentrate, change buffer and store as described in $C7$, above.

---

and, furthermore, the flow rates are poor because a large volume of a crude cell extract is applied to the column, which results in fouling of the column matrix. Loading can require 4−7 h, and elution is then performed overnight. The time required to complete subsequent ion-exchange fractionations becomes shorter as the protein is purified. A single redox protein will often elute from an ion-exchange column in two overlapping peaks. These peaks are due to the difference between the oxidized and reduced forms of the protein. Inclusion of 1 mM ascorbate in the buffer may prevent this anomalous resolution, but spectral analysis of a few fractions to confirm the diagnosis makes this unnecessary in routine preparations.

To determine the pore size of the column matrix to be used for size-exclusion chromatography, one can obtain an estimate of the molecular weight by performing sodium dodecyl sulphate polyacrylamide gel electrophoresis (SDS−PAGE) and detecting the target protein by using a specific test procedure. Denatured cytochromes c and some cytochromes b emit a red fluorescence upon transillumination with mid-range UV light (302 nm) (5). This can be used to visualize 100−200 pmol of these cytochromes, following SDS-PAGE of crude protein mixtures (the sensitivity is reduced if urea is present in the gel buffer). Alternatively they may be detected by staining for heme (6). Due to the low molecular weight of these cytochromes, concentrations of 15% acrylamide or a concentration gradient up to 20% are preferred for SDS-PAGE.

To prevent loss of resolution during size-exclusion chromatography, it is necessary to concentrate the samples to less than 5% of the column volume prior to loading. This is performed by ultrafiltration, using pressurized nitrogen. Nitrogen is used to prevent the oxidation damage that may be caused by pressurized air.

To maintain high resolution during size-exclusion chromatography, a buffer of high ionic strength is used to minimize inter-protein interactions. As this step is followed by ion-exchange chromatography, however, it is necessary to reduce the ionic strength, or change the buffer. In this case of cytochrome c_L, the alkaline buffer is diluted slowly with water to reduce the ionic strength. Cytochrome c_H, however, is exchanged into an acidic buffer prior to loading onto the CM-cellulose column. Proteins of less than 15 000 daltons, including these cytochromes c, will pass through most types of dialysis tubing. An ultrafiltration cartridge of small pore size is used, therefore, to prevent the loss of this small protein and to allow the exchange to be performed in an hour rather than overnight.

2.1.3 *Criteria of purity*

Three criteria for purity are used:

(i)   The observation of a single protein band corresponding to a cytochrome following SDS-PAGE (15% w/v acrylamide monomer). That this band is a cytochrome can be demonstrated by using the visualization techniques discussed in Section 2.1.2.

(ii)   The extinction coefficient at the characteristic absorption peak. This criterion can be used for any protein with a characteristic absorption spectrum. At a given characteristic wavelength the extinction coefficients for most groups of cytochromes are similar (see *Method Table 2*).

(iii)   The ratio of the α-peak absorption (see *Method Table 2*) divided by the absorption at 280 nm will be constant for a given cytochrome when it is fully reduced. This parameter will tend toward a maximum as the cytochrome is purified as other contaminating proteins that absorb at 280 nm are removed. It is typical for this value to exceed 0.8 for a purified, small and soluble cytochrome c. Because of the variation between proteins, this parameter is only useful for a cytochrome that has been purified and characterized previously.

The technique of proteolytic fragment mapping is a useful procedure that is used to determine if a purified protein, of similar properties to a larger protein, is in fact a fragment of this larger protein. This method has been used to demonstrate that the two small soluble cytochromes c of methylotrophic bacteria are distinct proteins (7). This was particularly important as cytochrome c_L is known to lose a 4000-dalton protein fragment during its purification. The purified proteins are subjected to partial proteolysis, and the fragments separated by SDS−PAGE. If the smaller protein is derived from the larger, many of the fragments obtained from the two proteins will have a similar mobility.

The purification of cytochrome c_H achieved is typically 200-fold with a yield of 20−40%. The last step may appear to reduce the purification (*Table 1*) but this step is often necessary to remove traces of cytochrome c_L which will otherwise contaminate the preparation. The purification of cytochrome c_L is usually in excess of 120-fold with a yield of 10−30% (*Table 1*). Absorption spectra of the reduced purified cytochrome c_H are presented in *Figure 1a*. The spectrum for cytochrome c_L is similar, but not identical.

2.2 Purification of the membrane-bound cytochrome *co* from *Methylophilus methylotrophus*

2.2.1 *Preliminary questions*

The development of a protocol for the purification of this membrane-bound cytochrome oxidase was fraught with problems. A preliminary characterization carried out according to *Method Table 1* showed that the only favourable characteristic was the availability

Table 1. Typical results for purification of the soluble cytochromes *c* from *Methylophilus methylotrophus*.

Purification step	Volume (ml)	Total cytochrome (nmol)	Cytochrome content (nmol/mg protein)	Yield (%)	Purification (fold)
Cytochrome c_H					
Soluble cell fraction	500	4690	0.6	100	1
Acid treatment	504	3800	1.4	81	2
DEAE-cellulose	487	2130	18.9	45	30
Sephadex G-150	92	1630	81.5	35	52
CM-cellulose	65	1300	169	28	264
Biogel P-10	35	910	134	19	209
Cytochrome c_L					
Soluble cell fraction	500	2100	0.3	100	1
Acid treatment	504	1740	0.6	83	2
DEAE-cellulose	483	957	28.1	46	98
Sephadex G-150	113	780	33.9	37	118
DEAE-cellulose	162	632	30.1	30	105
Sephadex G-75	32	623	36.6	30	128

The table describes the results obtained from a typical purification. Details are given in *Method Tables 3 and 4*. For the calculation of yields it is assumed that the ratio of the two cytochromes *c* recovered after the first DEAE-cellulose step is the same as the ratio in the initial soluble cell fraction; i.e. cytochrome c_H was 69% and cytochrome c_L 31% of the total cytochrome *c* in this fraction.

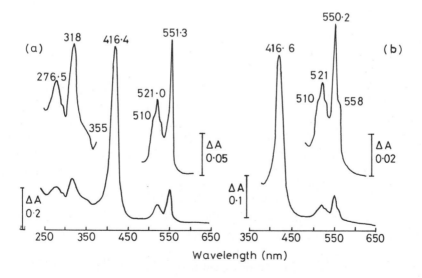

Figure 1. Absorption spectra of the reduced cytochrome c_H (**a**) and cytochrome *co* (**b**) of *Methylophilus methylotrophus*. The cytochromes were reduced with a few crystals of sodium dithionite. Spectra were measured at 20°C in (**a**) 5 mM Mops−NaOH buffer (pH 7.0) and (**b**) 12 mM Mops−NaOH buffer (pH 7.0) containing 0.2% (v/v) Triton X-100. The protein concentration was (**a**) 0.19 mg ml^{-1} and (**b**) 0.6 mg ml^{-1}. The light path was 10 mm.

of a rapid assay. The catalytic activity is not stable to even moderate changes in pH, irreversible loss occurring in a few hours if the pH is changed by as little as 0.5 of a pH unit from the optimum (pH 7.0). The activity decreases significantly after a few days' storage at 4°C. The enzyme cannot tolerate exposure to solutions of high ionic strength for more than a few minutes (for example 400 mM KCl, 75 mM $(NH_4)_2SO_4$). The enzyme is also unstable in the presence of a number of detergents. As a consequence of the extreme lability of this oxidase, the development of a protocol for its purification took over two years, and there remains a great deal of variation between batches. Similarly, attempts to purify the other cytochrome oxidase of this organism, cytochrome aa_3, have never been successful. Due to the low stability of the activity of the cytochrome *co*, it is necessary to complete the purification within 36 h of thawing the stored membrane fraction.

Since the catalytic activity of a cytochrome can be reduced or eliminated by a fractionation step without altering its spectral properties, it is necessary to perform enzyme assays for cytochromes with catalytic activities. The cytochrome *co* is routinely detected by measuring oxygen uptake in the presence of reduced N,N,N′,N′-tetra-methyl-*p*-phenylenediamine (TMPD). With practice, two Clarke-type oxygen electrodes can be used to assay samples at the rate of 8−12 per hour. Fortunately it is often not necessary to assay all of the samples obtained from a fractionation because those containing this cytochrome can be identified by their red coloration. If the concentration of enzyme is low this may not be possible, in which case this, and most other cytochromes, can be detected at concentrations below 1 μM by their strong absorption between 400 and 450 nm.

TMPD is not the physiological substrate. Furthermore, it is not clear if the presence of cytochrome *c* is a requirement of this assay. If it is, then it is no coincidence that all of the cytochromes that have been purified from a variety of bacteria using this assay contain at least trace amounts of endogenous cytochrome *c*. In order to eliminate this potential error during the development of this protocol, it was necessary to assay the pooled fractions from each purification step for their ability to oxidize horse heart cytochrome *c* and the two soluble cytochromes *c* purified from *Methylophilus methylotrophus*. The pattern and sensitivity of these pooled fractions to inhibition by cyanide and azide was also measured and compared to the properties of the enzyme in whole bacteria. The ability to bind carbon monoxide, a characteristic of certain cytochromes, was also monitored (see *Method Table 2*). The retention of the physiological activity was thus assured during the development of this protocol. These tests are, therefore, not required for subsequent routine purifications.

The organism *Methylophilus methylotrophus* produces maximal amounts of this cytochrome when grown on methanol under conditions of oxygen limitation. Further-more, the cytochrome aa_3, which has a similar catalytic activity and interferes with the assay for cytochrome *co*, is not produced under these conditions.

2.2.2 Protocol

The procedures are described in *Method Table 5* (8). The use of sonication to disrupt the bacteria results in a loss of enzyme activity, therefore the use of a French press is preferred.

The activity of this enzyme increases in the presence of the non-ionic detergent Triton

Method Table 5. Purification of the membrane-bound cytochrome *co* of *Methylophilus methylotrophus*.

Operations are performed at 4°C.

A. *Solubilization*
  1. Resuspend 1.2−2.4 g of protein from the membrane fraction (see *Method Table 3*) in 25 mM Mops−NaOH buffer (pH 7.0).
  2. While stirring gently add sequentially water, aqueous Triton X-100 (20% (v/v)) and 2 M MgCl$_2$, to give final concentrations of: 15 mg ml^{-1} protein, 12 mM Mops−NaOH buffer pH 7.0, 2.25% (v/v) Triton X-100 and 100 mM MgCl$_2$.
  3. After gentle agitation for 30 min, centrifuge the solution (135 000 *g*, 30 min) and discard the pellet.
  4. Dialyse the deep red supernatant for 1 h in an ultrafiltration cartridge (200 mol. wt cut-off) against 1 litre of 12 mM Mops−NaOH buffer containing 0.5% (v/v) Triton X-100 (pH 7.0).
  5. If necessary clarify by centrifugation as described in 3.

B. *Ion-exchange and size-exclusion chromatography*
  1. Apply the solubilized protein solution to a DEAE-cellulose (Whatman DE-52) column (5 × 5 cm) equilibrated with 12 mM Mops−NaOH buffer containing 0.5% (v/v) Triton X-100 (pH 7.0). Collect 10 ml fractions. The cytochrome *co* will not be adsorbed.
  2. Apply the eluate containing the non-adsorbed proteins to a CM-cellulose (Whatman CM-52) column (3 × 11 cm) equilibrated with 12 mM Mops−NaOH buffer containing 0.1% (v/v) Triton X-100 (pH 7.0). Collect 10 ml fractions. The cytochrome *co* will not be adsorbed.
  3. Concentrate to 30 ml while under nitrogen using a 50 000 mol. wt cut-off ultrafiltration membrane (Ulvac G-50-t).
  4. Apply in two portions to an upward flow Fractogel TSK HW 55 (S) size-exclusion column (2.4 × 80 cm, Merck) equilibrated with 12 mM Mops−NaOH buffer containing 0.2 (v/v) Triton X-100 (pH 7.0). Collect 3 ml fractions.
  5. Pool the fractions with the highest specific activity and store in liquid nitrogen.

X-100 but is lost irreversibly after treatment of the membrane fraction with the ionic detergent sodium deoxycholate, or the N-alkyl sulphobetaine zwitterionic detergent, Zwittergent SB 3-14. These observations are not uncommon for membrane proteins. Having selected Triton X-100 as the most suitable detergent, the variable parameters of the solubilization procedure have been optimized for both yield and subsequent stability of the oxidase activity. During such an optimization it is necessary to assay the enzyme at intervals of a few hours and one or more days, because many procedures that increase the yield for this solubilization step also cause a rapid irreversible loss of activity. The important parameters to be optimized are detergent concentration, protein concentration (and the ratio between the two), concentration of buffer and any salts added, temperature and time (9). During the development of the purification protocol, the effect of these parameters was assessed by treating the membrane fraction to the conditions to be tested, centrifuging (135 000 g, 30 min) and assaying the supernatant at timed intervals, dialysing where appropriate.

The values for the optimised parameters are given in *Method Table 5*. Higher detergent – protein ratios give an increase in the yield but result in decreases in the specific activity and stability of the activity for this enzyme. Higher concentrations of the buffer result in a decrease in the yield and stability of the enzyme, while lower buffer concentrations are unlikely to provide sufficient buffering capacity. Lower or higher concentrations of $MgCl_2$ result in a decrease in both the yield and the stability of the activity. Even in the procedure used routinely, it is necessary to remove the $MgCl_2$ within 90 min by rapid diafiltration, conventional dialysis being too slow to prevent unacceptable losses of activity. The extreme lability of this enzyme with respect to salt and changes of pH prevents the use of many techniques. For example, precipitation by acidification or by ammonium sulphate or acetate, or the use of preparative isoelectric focusing, all cause unacceptable losses of activity.

The activity of the solubilized enzyme decreased rapidly with time. It is necessary to complete all of the stages from the thawing of the stored membrane fraction to the application on to the size-exclusion column in a single working day. It is more conventional to have the size-exclusion fractionation step chronologically disposed between the ion-exchange columns. This was so in a prototype protocol but the sequence described in *Method Table 5* is as effective, and is used for reasons of speed. Because of the low tolerance of this enzyme to changes in pH, the ion-exchange chromatographic steps are performed at pH 7.0. Fortunately the major contaminating proteins are highly charged and are, therefore, removed by the ion-exchange matrices under this non-ideal condition. The decision to forego some loss of yield in return for a high specific activity during solubilization is also important in this respect.

A Fractogel size-exclusion matrix is used because it suffers less shrinkage in the presence of the detergent than many other matrices.

2.2.3 *Criteria of purity*

Three criteria of purity were used in the development programme; the third is used for routine preparations.

(i)   A single symmetrical protein band that oxidizes TMPD, following non-denaturing gel electrophoresis. The ability to oxidize TMPD is determined according to the method of deVrij *et al.* (10).

(ii)   A single symmetrical protein peak that oxidizes TMPD, following size-exclusion HPLC chromatography.

(iii)   Two protein peaks, at a ratio of approximately one, that contain cytochromes which fluoresce with a red colour, following SDS-PAGE (see Section 2.1.2).

The following procedure is used to indicate if two proteins are integral subunits of the same complex. SDS-PAGE is performed on sequential fractions obtained from the size-exclusion column. The ratio of any two subunits of the same protein complex should remain constant for each fraction and this has been found to be so for this cytochrome *co*. This procedure, however, has been criticized (11). It is necessary, therefore, to demonstrate that all fractions that contain only one of the two subunits also lack enzyme activity.

Table 2. Typical results for the purification of the cytochrome *co* from *Methylophilus methylotrophus*.

Purification step	Volume (ml)	Total activity ($\mu mol\ O_2\ min^{-1}$)	Specific activity ($nmol\ O_2\ s^{-1}\ mg^{-1}$)	Yield (%)	Purification (fold)
Membrane fraction	62	2108	14	100	1
Solubilized membrane	137	1532	21	73	2
DEAE-cellulose	194	1473	77	70	5
CM-cellulose	198	1456	119	69	8
Fractogel	44	635	211	30	15

The procedure is fully described in *Method Tables 3* and *4*, activities being measured with ascorbate/TMPD as substrate.

Because of the extreme lability of this enzyme, specific activity is not a reliable criterion of purity; SDS-PAGE is used routinely — see (iii) above. The detergent used, Triton X-100, has a large absorption peak at 276 nm which prevents the use of the ratio of the absorbance at the α-peak to that at 280 nm.

The purification of cytochrome *co* achieved is typically 15-fold with a yield of around 30% (*Table 2*). Absorption spectra of the reduced, (reduced) minus (oxidized), and (reduced plus carbon monoxide) minus (reduced) spectra of the purified cytochrome *co* are presented in *Figures 1* and *2*.

2.3 Purification of the transmembrane cytochrome oxidase from beef heart

2.3.1 *Preliminary questions*

The purification of the cytochrome oxidase from beef heart is a well-established procedure that has evolved over many years. The protocol presented here is part of a larger scheme in which all four respiratory complexes are purified from one batch of beef hearts. This protocol can be completed in 3 days.

Because the ionic detergents and the salts used in this protocol inhibit enzyme activity, the samples obtained from a fractionation step need to be dialysed prior to assay. This advantage is offset, however, by the relatively few samples obtained from procedures employing precipitation compared to those based upon column chromatography. Nonetheless, the assay used routinely is the spectrophotometric measurement of the heme *a* content (see *Method Table 2*) rather than the more time-consuming assays of catalytic activity.

2.3.2 *Protocol*

The procedures are described in *Method Tables 6* and *7* (12 — 15). This protocol is discussed below.

The complexity of the procedure for the isolation of a suitable source material for the purification of a mammalian protein (*Method Table 6*) is in strong contrast to the procedures used for bacterial cytochromes.

The relatively high concentrations of protein obtained at each stage allows the use of Biuret assay throughout the procedure. The advantage of this assay are speed, and a low susceptibility to interference from the variable components of the crude material

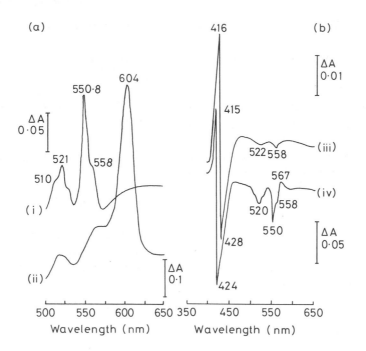

Figure 2. Difference spectra for the purified cytochrome oxidases. (**a**) (Dithionite reduced) minus (potassium ferricyanide oxidized) difference spectra for (i) cytochrome *co* from *Methylophilus methylotrophus*, and (ii) cytochrome oxidase from beef heart. (**b**) (Dithionite reduced plus carbon monoxide) minus (dithionite reduced) difference spectra of the cytochrome *co* of *Methylophilus methylotrophus* recorded after exposure to carbon monoxide for (iii) 30 sec (when the cytochrome *o* dominates the spectrum), and (iv) 30 min (when both cytochrome *o* and the more slowly reacting cytochrome *c* contribute to the spectrum). Spectra were recorded as described in *Method Table 2*, at 20°C with a 10 mm light path in 12 mM Mops−NaOH buffer (pH 7.0) containing 0.2% (v/v) Triton X-100, except for (ii) (0.1 M potassium phosphate buffer, pH 7.4). The protein concentrations were (i) 0.6, (ii) 1.0, and (iii,iv) 0.2 mg ml^{-1}.

Method Table 6. The preparation of beef heart mitochondria.

Operations are performed at 4°C.

1. (Day 1). Trim fat and connective tissue from heart. Mince the lean meat in a heavy duty industrial mincer. Weigh.
2. Suspend each kg of mince in 3 litres of 10 mM K_2PO_4 containing 0.25 M sucrose. Homogenize 0.8 litre aliquots in a Waring blender for 45 sec.
3. Add 6 M KOH to bring the pH to between 7.2 and 7.4.
4. Centrifuge (1600 g, 7 min) and pour the supernatant through muslin to remove any large particles of fat.
5. Dilute with 0.2 volumes of 0.25 M sucrose.
6. Centrifuge in a continuous-flow centrifuge at 65 000 g at a flow rate of 0.5 litre min^{-1}. Resuspend the brown mitochondria in 0.5 litres of 0.25 M sucrose and re-centrifuge. Repeat this process until the supernatant is no longer pink.
7. Resuspend in 0.25 M sucrose and determine protein (Biuret). Store in aliquots containing 6 g protein in 100 ml. Freeze at −20°C.

Method Table 7. The purification of cytochrome oxidase from beef heart mitochondria.

Operations are performed at 4°C.

A.   Pre-solubilization wash

1.   (Day 2). Thaw one aliquot of mitochondria (see *Method Table 6*) and dilute with 50 mM Tris−HCl buffer containing 0.67 M sucrose (pH 8.0) to give a final protein concentration of 23 mg ml^{-1}.
2.   While stirring continuously, add 2.73 ml of 100 g litre^{-1} sodium deoxycholate and 3.3 g of solid KCl per g of protein. Stir for a further 5 min.
3.   Centrifuge (7500 *g*, 30 min). Resuspend the green pellet in a minimal volume of 50 mM Tris−HCl buffer containing 0.67 M sucrose (pH 8.0). Store at −70°C until required.
4.   (Day 3). Thaw and resuspend in 0.1 M potassium phosphate buffer pH 7.4. Centrifuge (105 000 *g*, 15 min) and resuspend in the same buffer.
5.   Measure the protein content (Biuret) and adjust to 23 mg ml^{-1} protein using the same buffer.

B.   Solubilization and precipitation

1.   While stirring continuously add 20 g litre^{-1} potassium cholate to a final concentration of 1 mg per mg protein. Continue stirring.
2.   Make a saturated, aqueous ammonium sulphate solution and bring to pH 7.0 with NaOH. Add 1 ml of this solution to every 3 ml of protein solution, to bring the ammonium sulphate to 25% saturation. Stir for a further 5 min and then stand for 1 h.
3.   Centrifuge (75 000 *g*, 15 min). While stirring add 9 ml of neutralized, saturated ammonium sulphate solution to 100 ml of supernatant, to bring this to 31.2% saturation. Stand for 30 min.
4.   Centrifuge (75 000 *g*, 15 min). Decant the supernatant from the red/brown precipitate which is coated with a white film. While stirring add 8 ml of neutralized, saturated ammonium sulphate solution to 100 ml of supernatant, to bring this to 36.3% saturation. Stand for 20 min.
5.   Centrifuge (75 000 *g*, 15 min). While stirring add 12 ml of neutralized saturated ammonium sulphate solution to 100 ml of supernatant. Stand for 20 min.
6.   Centrifuge (75 000 *g*, 15 min). Resuspend the dark green precipitate in the minimum volume of 50 mM Tris−HCl buffer containing 0.67 M sucrose (pH 8.0) or 0.1 M potassium phosphate buffer pH 7.4, as required.
7.   Determine protein concentration (Lowry). Dilute to 25−40 mg ml^{-1} protein and store in 5−10 mg aliquots at −70°C.

---

and from the detergents employed. For a detailed method see (1), Chapter 1.

   The pre-solubilization wash with sodium deoxycholate serves to disrupt the mitochondrial membranes and acts as a fractionation step, removing lipids, other respiratory complexes and a number of other proteins. The protein is then transferred to a different buffer by centrifugation and resuspension.

To reduce the time required, the solubilization and first precipitation steps are performed without an intermediate centrifugation. Ammonium sulphate is used in this procedure, although for membrane proteins it is more usual to employ ammonium acetate. The latter is used in the longer scheme to purify the other respiratory complexes (13). When the lipid and/or ammonium sulphate concentration is high, the precipitates are sometimes obtained as pellets that 'float' on the supernatant. This is a common feature of precipitations carried out in the presence of detergents, and especially so in the case of the non-ionic detergents. For the purification of beef heart cytochrome oxidase ionic detergents are used because they can be removed easily, and this simplifies the reconstitution of the purified oxidase into lipid vesicles compared to the procedures necessary for this operation when non-ionic detergents are used. Furthermore, fractionation procedures using salt precipitation are usually more successful when performed in the presence of ionic detergents. The major disadvantage of the ionic detergents, irreversible inactivation, is not observed in this case.

Multiple precipitations are employed to ensure that the final preparation is of high purity. The initial fractionations remove lipids and proteins that contaminate the cytochrome oxidase preparation if this protocol is not adhered to rigorously.

It should be noted that the relative simplicity and speed of this purification, based upon precipitation by salt fractionation, compared to those discussed above that employ a variety of column matrices, has been obtained at the expense of a long period of development.

2.3.3 *Criteria of purity*

The heme *a* content per mg of protein (typically $7-12$ nmol mg^{-1}) is used as the criterion of purity (see *Method Table 2* and *Figure 2*). The large size of this complex produces a very small ratio of the absorbances of the α-peak/280 nm, making this criterion insensitive to contamination by other proteins. The subunit composition of the purified oxidase is usually confirmed by SDS-PAGE (*Figure 3*). The ability of the preparation to oxidize horse heart cytochrome *c* in the presence of ascorbate and TMPD is normally established only at the end of the procedure.

3. PURIFICATION OF GROWTH FACTORS—C.George-Nascimento and J.Fedor

A large number of polypeptide mitogens have been described since the discovery of Cohen (17) of epidermal growth factor (EGF) and by Levi-Montalcini and Hamburger (18) of nerve growth factor (NGF). They have been named for their target cell specificity—EGF, NFG, Fibroblast Growth Factor (FGF)—or for the tissue of their origin—platelet-derived growth factor (PDGF). They exhibit some effects in common, such as mitogenesis and the stimulation of DNA synthesis, and some unique action, such as stimulation of bone resorption by Transforming Growth Factor α (TGFα) and the stimulation of collagen production by TGFβ.

This section discusses the purification of several growth factors, excluding lymphokines. Detailed protocols for selected prototypes are intended to be illustrative rather than definitive. Some general points will be evident. Growth factors are characterized by their extremely low abundance in natural sources, usually less than 1 μg per litre. However, the majority are relatively stable to conditions likely to destroy

128

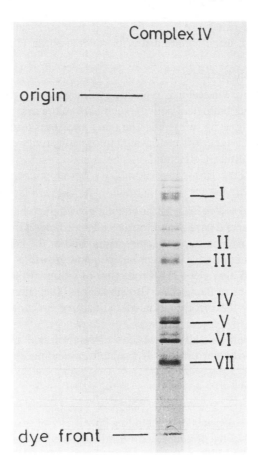

Figure 3. SDS-polyacrylamide gel electrophoresis of the purified cytochrome oxidase from beef heart. Beef heart cytochrome oxidase (Complex IV) was prepared by incubating 0.4 mg in 60 μl in 10 mM N-ethylmaleimide for 2 h at 0°C (to block disulphide exchange reactions), and then dissociated in 4 M urea containing 2.5% SDS for 2 h at room temperature. 15 μg of protein was loaded on to gradient slab gels (12.5−16% acrylamide) using the discontinuous buffer system of Laemmli (16), which contained 0.1% SDS throughout the system. The protein bands were visualised with Coomassie brilliant blue.

the activity of other proteins, for example, elevated temperatures or high concentration of organic solvents such as acetonitrile. Their relative robustness is thus responsible for the common use of HPLC techniques in their purification.

A fruitful general strategy has been to isolate sufficient amounts to obtain a partial amino acid sequence followed by cloning and expression using recombinant DNA techniques. Larger quantities can then be produced and formulated stably for pharmaceutical use. Therefore this section illustrates the purification of growth factors from both natural and recombinant sources.

Some growth factors bind to heparin; this specific binding is a key feature in any purification scheme of this class of growth factors. The biological role of heparin is not fully understood; however, it can be speculated that the affinity of growth factors for heparin is related to their function in direct growth and migration of cells *in vitro*.

129

The properties of other groups are not uniform, and their purification requires a variety of classical techniques, usually culminating in reverse-phase HPLC.

3.1 Assays for growth factors

The purification of a growth factor is typically evaluated with assays of two different types. One measures its biological activity such as a mitogenic response of cultured cells or specific binding to its receptors. The other measures purity, for example, by HPLC. Appropriate cell types for measuring biological activity can be obtained from the American Type Culture Collection.

3.1.1 *In-vitro bioactivity assays*

The majority of *in vitro* assays for heparin-binding growth factors use endothelial cells, but growth factors in general have a broad target cell specificity. Growth factors typically affect the density dependence, serum requirements, and/or the anchorage-dependence of cultured cells. Promotion of anchorage-independent growth is assayed by counting colonies formed in soft agar (19−21). Promotion of cell growth (22) and serum-free growth can be assayed by counting cells. Growth factors often stimulate DNA synthesis, which is measured by the incorporation of radioactive precursors such as tritiated thymidine.

However, all these assays take several days to perform and are therefore not very convenient for monitoring purification. If available, a binding assay will give quicker results.

3.1.2 *Binding assays*

Another class of assays depends on the binding specificity of the growth factor for its receptor, or for an antibody. Radioreceptor binding assays evaluate the ability of the preparation to bind to a specific receptor and to compete with an appropriate standard. The assay for EGF in *Method Table 8* is a prototype. Cell density is important in these assays because the number of EGF receptors is increased and their affinity is reduced as A431 cells reach confluence (23). Because cell density can be difficult to control, a series of standards are assayed.

Radioimmune assays (RIA) compare the ability of samples to compete with an iodinated standard for antibody binding, and sometimes rely on the cross-reactivity of the antibody to different growth factors (24,25). Western blotting (26), which also requires specific antibody, allows the detection of degraded species and gives structural information of the growth factor under study.

3.1.3 *HPLC assays*

Reverse-phase HPLC assays give the best resolution. A typical procedure is given in *Method Table 9*. Once the elution position of a growth factor is confirmed by bioactivity or binding assays, the method can be optimized by focusing on a part of the gradient.

3.2 Purification of non-heparin-binding growth factors

3.2.1 *Epidermal growth factor (EGF)*

This is a single polypeptide, $M_r \sim 6000$, which stimulates the proliferation of epithelial

Method Table 8. Radioreceptor assay for EGF and TGFα.

1. Grow A431 human epidermal carcinoma cells to confluency in Removacell microtitre plates (Dynatech Laboratories, Inc.) or similar plates.
2. Carefully wash 2× with room temperature PBS.
3. Add 8.2 ml 37% formaldehyde to 91.8 ml PBS to make 3% formalin solution, and dispense 100 μl into each well.
4. Incubate the plate at room temperature for 10 min.
5. Allowing a 5 min soak, wash 2× with PBS, 0.1% BSA.
6. Store up to 3 days at 4°C.
7. Add serial dilutions between 5−1500 ng ml^{-1} of either growth factor standard or sample in PBS, 0.1% BSA.
8. Dilute ^{125}I-mEGF (Amersham) 1:20 in PBS, and add 20 μl to each well.
9. Gently tap the plate to mix.
10. Incubate at 37°C for 1.5 h.
11. Wash the plate 2× with PBS, 0.1% BSA.
12. Transfer the individual wells to counting vials, and count in a gamma counter.

Method Table 9. HPLC assay for growth factors.

1. Equilibrate an analytical C18 column (4.6 mm × 15 cm) in 0.05% TFA in HPLC water at 1 ml min^{-1}. Set detector at 225 nm.
2. Clarify sample by 0.2 μm filtration or by centrifugation (e.g. 10 000 g, 5 min).
3. Inject 20−50 μl of the sample.
4. Develop with a linear gradient from 0−60% acetonitrile in 0.05% TFA over a period of 30 min.
5. Increase acetonitrile concentration to 80% and continue washing for 5 min.
6. Return to 0% acetonitrile over a period of 5 min.
7. Re-equilibrate for 10 min.

Method Table 10. Biorex 70 (Bio-Rad) chromatography of EGF.

1. Load an extract of homogenized tissue on to a column pre-equilibrated with 0.1 M acetic acid.
2. Wash the column with 4 vols of 0.1 M acetic acid.
3. Wash the column with 2 vols of 50% ethanol.
4. Elute the growth factor with 0.1 M HCl in 80% ethanol.
5. Titrate the eluate to neutral pH with dilute NaOH.
6. Remove the ethanol by rotary-evaporation.

and epidermal cells, and mediates several metabolic events such as DNA synthesis, Ca^{2+} transport and stimulation of glucose uptake (27). The protocol used by Savage and Cohen (28) to purify EGF from mouse submaxillary glands is the most popular of several purification schemes. Details are given in *Method Table 10*. For ion-exchange chromatography, Bio-Rad 70 (Bio-Rad) should be used as it has a higher binding capacity than DEAE-Cellulose. After ion-exchange chromatography, EGF is readily concentrated by ultrafiltration, in preparation for gel filtration. Ultrafiltration is fast, and easy to scale-up using commercial spiral cartridges (Amicon, Millipore). Good recovery is likely with a 2000 molecular weight cut-off membrane. Gel filtration on a Bio-Gel P-10 (Bio-Rad) or less rigid resins, and a second ultrafiltration prepares the growth factor for HPLC.

Reverse-phase HPLC on C18, C8 or C4 columns is appropriate for this hydrophobic protein. EGF is fairly stable when eluted with a gradient of acetonitrile in a mobile phase containing trifluoroacetic acid (TFA). Lyophilizing the eluate results in a powder containing EGF as a trifluoroacetate salt. Urinary EGF can be purified by the method of Gregory and Willshire (29).

3.2.2 *Transforming growth factor alpha (TGFα)*

This growth factor is purified from serum-free medium conditioned by Snyder-Theilen feline sarcoma virus-transformed Fisher rat embryo cells (30), or by Moloney murine sarcoma virus-transformed 3T3 cells (31). TGFα has 20−28% sequence homology with EGF (30), binds to the EGF receptor of A431 cells, and causes down-regulation with kinetics similar to that caused by EGF (32).

Purification of TGFα from natural sources is difficult (*Method Table 11*) because it must be separated from other growth factors. Derynck *et al.* (33) have purified recombinant TGFα from *E. coli*, either as a fusion protein or as mature TGFα. In the

Method Table 11. Purification of TGFα.

---

1. Collect conditioned medium and add 0.1 M phenylmethanesulphonyl fluoride (Sigma) in isopropanol to a final concentration of 1 mM.
2. Filter out detached cells and concentrate with a DC10 concentrator using a 100 000 molecular weight cut-off hollow fibre cartridge (Amicon).
3. Clarify the concentrate by re-filtering with a 5000 molecular weight cut-off cartridge.
4. Dialyse the concentrate for 50 h at 4°C in 3500 molecular weight cut-off Spectrapore 3 tubing (Spectrum Medical Industries) against 100 vols of 0.1 M acetic acid.
5. Centrifuge for 1 h at 30 000 g.
6. Lyophilize the supernatant, and dissolve with 1 M acetic acid to 25 mg ml^{-1}.
7. Clarify by centrifuging at 100 000 g for 45 min.
8. Load 6 ml on to a 2.5 × 100 cm column of Bio-Gel P60 (200−400 mesh) (Bio-Rad) and elute at 12 ml h^{-1} with 1 M acetic acid.
9. Lyophilize the fractions containing soft-agar growth promoting activity, and redissolve them in 0.6% trifluoroacetic acid−acetonitrile or 0.3% trifluoroacetic acid−1-propanol.
10. Fractionate the pooled gel filtration fractions by reverse-phase HPLC on μBondapak C18 columns.

---

latter case, the protein formed inclusion bodies which required solubilization in 7 M guanidine hydrochloride, followed by refolding at pH 9 in the presence of 1.25 M reduced glutathione/0.25 M oxidized glutathione. Methods for preparation of inclusion bodies can be found in Chapter 6.

3.2.3 *Transforming growth factor beta (TGFβ)*

This growth factor can either stimulate or inhibit proliferation, differentiation and cell type-specific processes. The signals generated from the TGFβ receptor can block the action of other growth factors (35). The type of serum (plasma-derived or whole blood-derived) used in assays of soft agar growth can make a difference, because other platelet factors (36), as well as EGF and TGFα, modulate TGFβ action.

TGFβ can be isolated as a dimer of 23 000 molecular weight from the medium conditioned by Snyder-Theilen feline sarcoma virus-transformed rat kidney cells, which is also a source of TGFα. Massague (37) found a discrepancy between the relative recoveries of EGF receptor binding activity and colony-forming activity from the gel filtration step, and suggested that various factors which act coordinately were present in the medium. When the fractions of this gel filtration step are co-assayed with TGFα from the 6000 – 10 000-molecular-weight fractions or with EGF, the presence of TGFβ in the 14 000-molecular-weight fraction is revealed. TGFβ is purified further (38) by reverse-phase HPLC on μBondapak C18, chromatography on I-125 molecular filtration columns (Millipore-Waters) and elution from non-reducing SDS polyacrylamide gels.

TGFβ can also be purified from medium conditioned by Moloney murine sarcoma virus-transformed 3T3 cells (19), platelets (36), bovine kidney (34) or placenta (39). The overall yield is generally low (for example 0.1 – 0.5 μg per litre from conditioned medium or 40 μg from 13 kg kidney).

3.2.4 *Platelet-derived growth factor (PDGF)*

The mitogenic and other activities of PDGF are reviewed by Heldin *et al.* and Ross *et al.* (40,41). It is a difficult growth factor to purify because of its hydrophobicity. PDGF is a basic heterodimer of 30 000 daltons. It is resistant to heat and chemical denaturants, probably due to its many disulphide bonds. Loss of material can be reduced by using plastic or silicon-treated glassware.

The procedure of Heldin *et al.* (42) uses platelets disintegrated by repeated freezing and thawing, dialysis of the supernatant against 0.08 M NaCl, batch CM Sephadex chromatography, Blue Sepharose (immobilized Cibacron blue F3GA, Pharmacia) chromatography and elution with 50% ethylene glycol, BioGel P-150 chromatography in 1 M acetic acid and Sephadex G200 chromatography in 1 M acetic acid. An alternative procedure (41) uses heparin-Sepharose and phenyl-Sepharose in addition to cation exchange chromatography and gel filtration. Collaborative Research Incorporated supplies anti-PDGF for immunoaffinity chromatography.

3.2.5 *Tumour necrosis factor (TNF)*

The stimulation of leukocytes (and possibly other cell types) with mitogens results in the secretion of proteins called cytokines collectively known as Tumour Necrosis Factors. Mitogen-stimulated macrophages produce TNFα (also called lymphotoxin) and mitogen-stimulated lymphocytes produce TNFβ (also called cachectin) (43). TNFα and TFNβ are 50% homologous, and bind to the same cell surface receptor (45). Both recombinant TNF and non-recombinant TNF cause hemorrhagic necrosis of mouse tumours, are cytotoxic, and synergize with interferon (44).

The purification of TNF is monitored with a mouse L fibroblast cell lytic assay that involves treatment of cells with TNF in the presence of actinomycin D, and detection of lysed cells by staining with crystal violet in methanol−water. The relative 100% is standardized with lysis by 3 M guanidine hydrochloride. TNF is composed of three subunits of 17 000 molecular weight and is inactivated by the reduction of disulphide bonds.

Production of TNF in the HL-60 promyelocytic leukaemia cell line is induced by treatment with 10 ng ml^{-1} of 48-phorbol 12-myristate 13-acetate for 16−24 h. The cell-free filtrate is batch adsorbed on to Controlled-Pore Glass beads (Electro-Nucleonics) at 4°C (100 ml beads per five litre filtrate), poured into a column, washed with 1 M NaCl, 10 mM sodium phosphate pH 6.0, and eluted with 20% ethylene glycol, 1 M NaCl, 10 mM sodium phosphate pH 8. The eluate is loaded on to a DEAE-53 column (Whatman) in 0.01% Tween 20 (Sigma), 10 mM sodium phosphate pH 8, and eluted with 75 mM NaCl. The eluate is dialysed against 0.01% Tween 20, 1 mM azide, 20 mM Tris−HCl pH 8, concentrated, and applied to a Mono-Q column (Pharmacia) equilibrated in dialysis buffer and eluted in the 50−65 mM fractions of a 40−75 mM NaCl gradient. Subsequent reverse-phase HPLC on Synchropak RP-4 (Waters) with 0.1% trifluoroacetic acid and 1-propanol reduced biological activity (46).

Recombinant TNF has been purified from *E. coli* (47) and yeast (G.Kuo, Chiron Corporation, Pers. Comm.). In the case of *E. coli* the supernatant of a cell lysate was applied to DEAE-Sepharose CL-6B (Pharmacia), washed with 100 mM Tris−HCl pH 8, and eluted with 100 mM NaCl, 20 mM Tris−HCl pH 8. The fractions with cytolytic activity were applied to DEAE-Sepharose CL-6B and washed with 150 mM NaCl, 20 mM Tris−HCl pH 8. The active fractions in the flow-through were applied to Sephacryl S200 (Pharmacia) in 150 mM NaCl, 5 mM sodium phosphate pH 7.4, from which the activity eluted at a molecular weight of 45 000.

3.3 Purification of heparin-binding growth factors

The heparin-binding growth factors, including fibroblast growth factors (FGF), endothelial cell growth factor (ECGF), eye-derived growth factor (EDGF), retina-derived growth factor (RDGF), cartilage-derived growth factor (CDGF) and others, exhibit sequence homology, despite their diverse sources and target tissues. For a review, see Lobb *et al.* (48). Extraction, often from conditioned medium (*Method Table 12*), in an active conformation is critical. These factors elute from heparin columns under characteristic salt and pH conditions. Heparin may contaminate growth factor preparations by leaching from the affinity column and can itself modulate the migration of endothelial cells (49), or alter the activity of the growth factor (50). Purification thus typically involves cation exchange (*Method Tables 13* and *14*) or gel filtration chromatography (*Method Table 15*) after heparin chromatography (*Method Tables 16* and *17*).

Method Table 12. Obtaining conditioned medium for growth factor isolation.

1. Grow cells to confluence in medium with serum.
2. Rinse with phosphate buffered saline (PBS) three times.
3. Incubate for 3 days in medium without serum.
4. Collect the medium.
5. Clarify the medium by centrifugation or filtration.
6. Filter-sterilize the medium and store in polypropylene tubes at $-20°C$.

Method Table 13. CM Sephadex C50 (Pharmacia) chromatography for anionic growth factors.

1. Homogenize the appropriate tissue in 0.15 M ammonium sulphate, using 1 litre kg^{-1}.
2. Titrate the mixture to pH 4.5, and magnetically stir for $1-2$ h at $4°C$.
3. Titrate the mixture to pH 7, and add $40-80\%$ ammonium sulphate dropwise.
4. Centrifuge at 10 000 g for 30 min.
5. Redissolve the pellet in water, and dialyse against 0.1 M sodium phosphate pH 6.
6. Apply to CM-Sephadex C50 pre-equilibrated with 0.1 M sodium phosphate pH 6.
7. Wash the column with 0.15 M NaCl, 0.1 M sodium phosphate pH 6.
8. Elute with a NaCl gradient or with 0.5 M NaCl, 0.1 M sodium phosphate.
9. Eluate may be applied directly to a heparin column.

Method Table 14. Biorex 70 (Bio-Rad) cation exchange chromatography.

1. Apply 50 ml of tissue supernatant in 0.1 M Tris−HCl, 0.1 M NaCl pH 7 to a 2.5 × 20 cm column of Biorex 70 (Bio-Rad), $200-400$ mesh.
2. Wash with 50 ml of the same buffer.
3. Elute with a linear gradient of $0.1-1.0$ M NaCl at a flow rate of 30 ml h^{-1}.
4. Eluate may be applied directly to a heparin column in the presence of as much as 0.6 M NaCl.

Method Table 15. Sephacryl S200 (Pharmacia) chromatography of heparin-Sepharose purified material.

1. Dialyse heparin-Sepharose eluate containing growth factor activity against 0.2 M NaCl, Tris−HCl pH 7.
2. Concentrate by ultrafiltration on Amicon P10 or YM10 membrane.
3. Apply 1 ml onto a 1 × 90 cm column pre-equilibrated with the same buffer.
4. Elute at a flow rate of 1 ml h^{-1}.

Method Table 16. Batchwise heparin-Sepharose purification of crude tissue extracts.

1. In a Waring blender, homogenize the tissue in 1.3 vols of 50 mM Tris-HCl pH 7.4, 50 mM EDTA (Tris-HCl, EDTA).
2. Centrifuge at 10 000 g for 30 min.
3. Titrate the supernatant to pH 4.5 with 1 M acetic acid.
4. After stirring the supernatant for 30 min, centrifuge at 10 000 g for 30 min.
5. Titrate the supernatant to pH 7.4 with 1 M NaOH.
6. Add saturated ammonium sulphate to 50%.
7. Centrifuge at 10 000 g for 30 min.
8. To the supernatant, add saturated ammonium sulphate to 95%.
9. Centrifuge at 10 000 g for 30 min.
10. Resuspend the pellet in Tris-HCl, EDTA using 10 ml for each kg of tissue used.
11. Using dialysis tubing with a low molecular weight cut-off, dialyse the suspension against 50 vols of Tris-HCl, EDTA.
12. Clarify the suspension by centrifuging at 10 000 g for 1 h.
13. To the supernatant, add 20 ml hydrated heparin-Sepharose (Pharmacia) for each kg of tissue used.
14. Gently rock or gyrate the resulting slurry for 1 h.
15. Pack the material into a column with 2.5 cm inner diameter.
16. Wash the column with 10 vols Tris-HCl, EDTA, then with 1 vol 0.5 M NaCl at a flow rate of one vol per h.
17. Elute the bound proteins with a gradient of 0.5 M−2.0 M NaCl in Tris-HCl, EDTA, using a total of 15 vols.

Method Table 17. Heparin-Sepharose chromatography of partially purified material.

1. Equilibrate a 1 × 8 cm column with 0.1 M NaCl, 0.01 M Tris-HCl pH 7.
2. Dissolve 600−1000 units of 20 ml of growth factor in this buffer and apply it to the column.
3. Wash the column with 20 ml of buffer.
4. Elute the bound protein with 300 ml of a gradient of 0.1−3.0 M NaCl, 0.1 M Tris-HCl at a flow rate of 30 ml h$^{-1.}$

3.3.1 *Heparin chromatography*

Using an affinity technique requires choosing one of two methods:

(i) a batch process, where the starting material is mixed with the affinity-modified matrix and subsequently packed into a column (*Method Table 8*); or

(ii) loading of the starting material on to a prepacked column (*Method Table 8*).

The batch process allows more time for binding prior to elution. This is useful when the growth factor is only a small fraction of the total protein in the starting material (for example in crude material). The prepacked column can be used repeatedly with the same results, provided that leaching of the affinity agent from the matrix does not damage the binding capacity of the column. The prepacked column is especially convenient when all of the growth factor can be bound in one pass of the starting material through the column. Either Pharmacia's Heparin-Sepharose or Pierce's Heparin on Agarose should be used.

3.3.2 *HPLC of heparin-binding growth factors*

A low-pH reverse-phase HPLC is often used as a final purification step with a mobile phase containing 0.05% TFA and an acetonitrile or propanol gradient. Some growth factors are unstable at low pH, therefore their biological activity should be carefully monitored. For the solid phase, test the performance and reproducibility of C_{18}, C_8 and C_4 columns (Vydac, Millipore-Waters), 5–25 cm in length. A linear gradient of acetonitrile 22–50% over 15–45 min in 0.05% TFA at a flow rate of 0.6–1.7 ml min^{-1} should be tried, and, if necessary, further optimized. For analytical purposes use a 5 μm particle size with a 300 Å pore size, whilst for preparative purposes a larger particle size is more suitable.

3.4 Conclusion

This section provides basic protocols and general strategies for the purification of heparin binding growth factors and specific non-heparin binding growth factors. Growth factors often exhibit microheterogeneity, as discussed below for EGF. Human EGF (α-urogastrone) undergoes a post-translational modification which removes the carboxy-terminal residue (51). EGF's single methionine residue is susceptible to oxidation; the authors have found the sulphoxide derivative to be as active as the reduced form. The Gly^{15}–His^{16} bond of basic FGF is susceptible to cleavage, which may be either an artefact of the extraction process or the result of tissue-specific processing (52). The final product must therefore be carefully characterized by amino acid analysis and amino-terminal sequencing.

4. PURIFICATION OF COLLAGEN TYPES FROM VARIOUS TISSUES— V.C.Duance

Collagen is an unusual protein since it is normally insoluble, due to numerous cross-links between individual protein chains. Thus, one of the important stages of collagen purification is solubilization; this is achieved by proteolytic digestion. Alternatively, the precursor of collagen, procollagen, is isolated using neutral pH buffers. During the past decade a number of new collagen types have been discovered, currently numbering 13. It has become apparent that tissues once thought to be composed of a single type of collagen, e.g. tendon (type I) or cartilage (type II) are indeed very complex. For instance, five different collagen types have been isolated from cartilage and skin is known to contain at least six collagen types.

With such complexity, quite rigorous methods of purification need to be employed. This section describes a preparative procedure for the purification of types I and III procollagen from skin and types I, III, IV, V and VI collagen from human placenta, a common tissue source of these molecules. In addition, the purification of the constituent

137

α-chain components of type I collagen is briefly outlined. More detailed information on these collagen types and methods used can be found in numerous recent reviews (53–56).

4.1 Solubilization

4.1.1 *Procollagen*

For isolation of pro-types I and III collagen, foetal or new-born calf or rat skin is normally used (57,58). Yields of pro-type I collagen are low, as this molecule is processed by the tissue very quickly, but reasonable quantities of pro-type III can be obtained.

Procollagen, unlike collagen, is not cross-linked and is therefore readily extracted from tissues using neutral pH buffers. However, stringent precautions are required to prevent proteolytic digestion by endogenous tissue proteases.

(i)   Finely chop the tissue and homogenize in a Polytron homogenizer at 4°C with 10 volumes of 50 mM Tris−HCl, pH 7.4, 150 mM sodium chloride containing the following protease inhibitors: 1 mM phenylmethyl sulphonyl fluoride (PMSF), 1 mM sodium iodoacetate, 10 mM α-aminohexanoic acid, 5 mM benzamidine 1 mg litre^{-1} soyabean trypsin inhibitor and 10 mM EDTA.

(ii)   Incubate for 2 h at 4°C.

(iii)   Clarify by filtering through muslin and Celite. (The next stage of this purification is described in Section 4.2.1).

4.1.2 *Collagen*

Since most of the collagenous components of connective tissues are relatively insoluble, partial proteolytic digestion is required for their extraction. This also results in degradation of many of the contaminating proteins.

(i)   Homogenize the tissue in 10−20 vol of 0.5 M acetic acid using a Polytron homogenizer.

(ii)   Add pepsin to a ratio of pepsin: wet tissue of 1:100 w/w, and incubate at 4°C for 16 h.

(iii)   Remove the insoluble residue by centrifuging at 10 000 *g* for 30 min.

(iv)   The insoluble residue may be re-extracted as above to increase the yield. (The next stage of this purification is described in Section 4.2.2.)

Some tissues, such as cartilage, require pre-treatment of the tissue to remove proteoglycans before the pepsin can effectively solubilize the collagen (59). The degree of solubilization of the tissue is dependent to some extent on the age of the animal used; in general, the collagenous components of most tissues become less soluble with age.

Several of the more recently described collagens are partially degraded by the pepsin extraction procedure; the isolation of these molecules has been achieved from specific tumours in which the collagen cross-linking has been inhibited by a lathyrogen (60,61). This enables the molecules to be extracted in neutral pH buffers similar to that outlined above for procollagen.

4.2 **Purification**

4.2.1 *Types I and III procollagen*

The extraction procedure detailed above will solubilize un-crossed-linked types I and III collagens as well as the precursor molecules, pro-type I and pro-type III collagen. These can be purified by the methods of (57) and (58). In order to minimize proteolytic digestion, all procedures should be carried out at 4°C in the presence of protease inhibitors.

(i)   Add ammonium sulphate to 30% saturation to precipitate the collagenous proteins. Centrifuge at 15 000 g for 20 min at 4°C.

(ii)   Wash the precipitate twice with the extraction buffer to which sodium chloride has been added to 20%. Discard the soluble material.

(iii)   Extract the precipitate twice with the extraction buffer containing 1 M NaCl. Clarify the extract by centrifuging at 70 000 g for 15 min.

(iv)   Add solid NaCl to 20% w/v to precipitate the collagenous proteins. Centrifuge at 70 000 g for 15 min at 4°C.

(v)   Redissolve the pellet in the minimal volume of distilled water, and dialyse against 50 mM Tris−HCl, 200 mM NaCl, pH 7.6.

(vi)   Apply the dialysate to a DEAE-cellulose column (4 × 25 cm) equilibrated in the same buffer used for the dialysis. The collagenous proteins will pass through the column, whilst the acidic glycoproteins and proteoglycans will be retained.

(vii)   Dialyse the flow-through from the column against 50 mM Tris−HCl, 20 mM NaCl, 2 M urea, pH 7.5.

(viii) Apply the dialysate to a second DEAE-cellulose column (2.5 × 10 cm) equilibrated in the same buffer used for the dialysis. The types I and III collagens will not be retained by this column.

(ix)   Elute the types I and III procollagens with a 600 ml gradient from 20 to 200 mM NaCl. Monitor the eluate at 230 nm, since collagen has a low absorbance at 280 nm. Types I and III procollagen elute at approximately 50 mM and 100 mM NaCl respectively.

(x)   Add pepstatin to 1 μg ml^{-1} to pooled fractions and dialyse against 0.05 M acetic acid.

(xi)   Lyophilize and store desiccated at −20°C or below.

The unbound material from the second DEAE cellulose column contains types I and III collagen. These can be separated using the salt fractionation protocol detailed below for pepsin-extracted materials.

4.2.2 *Pepsin-extracted material*

Placenta is a common tissue source used for the preparation of types I, III, IV, V and VI collagen; the protocol detailed below, however, would be suitable for many other tissues.

Due to its triple-helical structure, a large proportion of the collagen molecule is resistant to pepsin digestion at 4°C, whereas most other tissue constituents are substantially degraded. Using salt fractionation of the pepsin extracts, solutions containing essentially pure collagen can be obtained. Subsequent anion-exchange chromatography is used to remove any remaining acidic glycoproteins and proteoglycans.

Figure 4. Preparation of pepsin extracted collagens from human placenta.

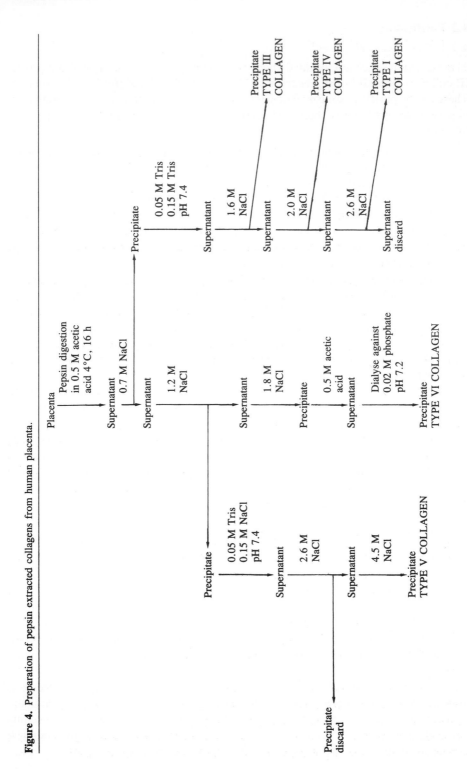

Table 3. Collagen precipitation at different salt concentrations.

Collagen type	Concentration of NaCl needed to precipitate	
	From 0.5 M acetic acid	From 0.05 M Tris pH 7.4
Type I	0.7	2.6
Type III	0.7	1.6
Type IV	0.7	2.0
Type V	1.2	4.5
Type VI	1.8	−

Collagens are fractionated initially from the 0.5 M acetic acid extract by addition of solid NaCl to give sequentially 0.7 M, 1.2 M and finally 1.8 M. The flow diagram in *Figure 4* illustrates this procedure and the salt concentrations at which each collagen type precipitates from acid and neutral pH buffers are shown in *Table 3*.

(i)   Add solid NaCl to give the 0.7 M final concentration.

(ii)   Incubate for several hours at 4°C, and then centrifuge at 50 000 *g* for 30 min.

(iii) Adjust the NaCl concentration of the supernatant to 1.2 M NaCl and repeat step (ii). Then adjust the NaCl concentration of the supernatant to 1.8 M NaCl and repeat step (ii).

(iv)   Dissolve the precipitates from the 0.7 M and 1.2 M extractions separately in 50 mM Tris−HCl, 150 mM NaCl, pH 7.4.

(v)   Add NaCl to give a final concentration of 1.6 M.

(vi)   Centrifuge at 50 000 *g* for 30 min. Redissolve the pellet in 50 mM Tris−HCl, 150 mM NaCl, pH 7.4.

(vii) Adjust the NaCl concentration of the supernatant to 2.0 M and repeat step (vi).

(viii) Repeat steps (vii) and (vi) sequentially adjusting the NaCl concentrations of 2.6 M and 4.5 M.

(ix)   Dissolve the pellet from the 1.8 M NaCl extract obtained in (iii) in 0.5 M acetic acid. Dialyse against 20 mM sodium phosphate buffer pH 7.2. Collect the precipitate by centrifuging at 50 000 *g* for 30 min.

DEAE-cellulose chromatography in 50 mM Tris chloride, 200 mM NaCl pH 7.6 is used to remove the acidic contaminants of types I, III and IV collagen.

Modifications of this procedure have been used for types V and VI collagen (62−64). The purity of collagen preparations is usually assessed by SDS−PAGE using the Laemmli system with an 8% separating gel and a 4% stacking gel (see Chapter 1 of ref. 1).

With many more collagen types being isolated from a wide variety of tissues, many variations of the above scheme exist. For more information on the different collagen types and methods for their preparation, see refs. 54, 55 and 65.

4.2.3 *Separation of α-, β- and γ-components*

Each collagen molecule is composed of 3 α-chains. Type I collagen contains 2 distinct α-chains, α1 and α2, and has a molecular composition $[\alpha 1(I)]_2 \alpha 2(I)$.

Collagen exists in the tissue as a highly cross-linked matrix. When it is extracted with pepsin, various-sized polymers are solubilized. On denaturation, covalently cross-linked α-chains are present as dimers (β-components), trimers (γ-components) and higher molecular weight aggregates. These components can arise by inter- or intramolecular cross-linking (66).

To separate the α-, β- and γ-components, size-exclusion chromatography is employed (67).

(i)   Equilibrate a column (5 × 150 cm) of BioGel A1.5 M or A5 M in 50 mM Tris−HCl, 1 M CaCl$_2$ pH 7.5.

(ii)   Dissolve the samples (up to 500 mg) in equilibration buffer and heat to 45°C for 15 min to denature the collagen.

(iii)   Centrifuge at 50 000 *g* for 30 min.

(iv)   Load the supernatant on to the column. Monitor the eluate at 230 nm. The α- and β-components should be well resolved, but the γ-component may be contaminated with higher molecular weight material.

(v)   Dialyse pooled fractions against 0.5 M acetic acid and lyophilize.

4.2.4 *Separation of α$_1$ and α$_2$ components*

To separate the α$_1$ and α$_2$ chains of type I collagen, CM-cellulose chromatography is normally used.

(i)   Equilibrate a CM-cellulose column in 40 mM sodium acetate, 20 mM NaCl, pH 4.8 at 45°C using a water-jacketed column. The buffer must be thoroughly degassed prior to use, to prevent bubble formation.

(ii)   Dissolve the sample in equilibration buffer and heat to 45°C prior to loading on to the column.

(iii)   Elute the column with a 20 mM to 200 mM NaCl gradient and monitor the eluate at 230 nm. The α$_1$ chain will elute prior to the α$_2$ chain. Any β- and γ-components will elute between the two α-components.

CM-cellulose chromatography has also been used for the separation of the constituent α-chains of other collagen types (63,68,69) as well as for the separation of native collagen fragments of type IX collagen (70,71). In addition, DEAE-cellulose chromatography has been used for the purification of the α-chain components of some collagen types (62).

5. PURIFICATION OF CONNECTIVE TISSUE METALLOPROTEINASES— G.Murphy

The metalloproteinases collagenase, stromelysin and gelatinase are synthesized and secreted by connective tissue cells and are thought to be responsible for the turnover of the macromolecular components of connective tissue extracellular matrices. They are Zn^{2+} metalloenzymes and require Ca^{2+} for full activity. Recent studies have shown that they have related primary structures and different, but overlapping, substrate specificities, which are common to the enzymes derived from different connective tissue cells, as well as from different mammalian species. Collagenase is the most specific, cleaving native fibrillar collagens (types I, II and III) at a single locus. Stromelysin degrades many matrix components such as the protein core of proteoglycan, the non-helical regions of type IV collagen, laminin, fibronectin and elastin. Gelatinase degrades the denatured forms of all collagens (gelatin) as well as native type IV and V collagens and elastin.

These enzymes are often produced co-ordinately by tissue and cells in culture and may act co-ordinately *in vivo* in some instances. They are secreted in a proform which is inactive, but can be activated with a concomitant fall in M_r of about 10 000 by treatment with organomercurials or proteinases. During purification these enzymes have a tendency to 'auto-activate', generating lower M_r products which are not all enzymatically active, but are immunologically detectable. Since the structure and properties of these enzymes are similar, their separation is quite difficult, and no single technique will definitively separate them. The protocol described here is one of several purification procedures available (72−80). However, the general strategy employed is similar in each case.

(i)   Concentration and removal of interfering culture medium components and glycos-aminoglycans;
(ii)   differential ion exchange chromatography;
(iii)   affinity or dye-ligand chromatography;
(iv)   gel filtration.

A flow diagram of the separation of collagenase, stromelysin and gelatinase is given in *Figure 5*. All procedures should be carried out at 4°C. Purified enzymes should be stored at −70°C.

5.1 Initial considerations

The metalloproteinases are purified in the latent form whenever possible. Active forms continously self-degrade to smaller fragments with different elution properties which aggravate the problems of overlapping chromatographic behaviour mentioned above. It is also known that stromelysin potentiates collagenase activation (82). Partial contamination of collagenase with stromelysin at the end of most purification procedures probably accounts for the enormous range of specific activities reported for this enzyme (73,79).

Enzyme activities are assayed by standard procedures at each stage of the purification and related to protein estimations made either by A_{280} measurements or by fluorimetric analysis (83).

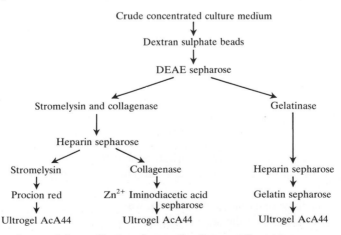

Figure 5. General protocol for purification of connective tissue metalloproteinases.

143

Tris buffers should be avoided, and cacodylate or borate used instead. The enzymes are converted to their active form by exposure to an organomercurial (4-aminophenyl-mercuric acetate) or by brief trypsin treatment (74). It is useful to compare the ratio of specific activities of the material at each stage, with respect to the different substrates: for example, collagenase specifically degrades type I collagen, but has very low activity either on gelatin, the preferred gelatinase substrate, or on casein which is most efficiently degraded by stromelysin. As the other enzyme activities are removed, the ratio of collagen − gelatin and collagen − casein degradation by the collagenase-containing pools will stabilize.

Purity of the proteins can be assessed by conventional SDS-PAGE analysis (84) and silver staining (85). The use of immunoblotting is also recommended for identification of breakdown products (86).

DEAE-Sepharose, heparin Sepharose, iminodiacetic acid Sepharose and Procion Red (Reactive Red Sepharose) can be obtained from Pharmacia. Cross-linked dextran sulphate beads and gelatin-agarose are available from Sigma.

5.2 Concentration of culture medium

The enzymes are generally harvested in very dilute solution (1 μg ml^{-1}) in the serum-free culture supernatants of connective tissue fragments or cell monolayers, previously enzymically dissociated or grown out from the tissue. Frequently, up to 10 litres of medium may be used and should be concentrated by hollow fibre or membrane ultra-filtration with a cut-off of M_r 5000. Buffering of the culture medium and addition of calcium ions and non-ionic detergent is recommended (final concentrations 20 mM Tris − HCl, pH 7.5, 10 mM $CaCl_2$, 0.05% Brij 35; further referred to as Buffer A).

An alternative or additional concentration may be effected by using dextran sulphate-Sepharose or cross-linked dextran sulphate beads.

(i)   Equilibrate a 200-ml column of the chosen matrix with Buffer A containing 0.15 M NaCl.
(ii)   Load the buffered medium equivalent to 5 litres containing about 3 g total protein.
(iii) Wash with Buffer A.
(iv)   Elute with Buffer A containing 2 M NaCl.

This step has the advantage of removing interfering glycosaminoglycans and a large amount of contaminating protein (*Table 4*). Furthermore, Phenol Red is also removed, since it does not interact with the matrix. The amount of total protein remaining after this step is of the order of 100 mg and subsequent chromatographic steps can be performed on smaller volumes of matrix media.

5.3 Separation of gelatinase

5.3.1 *DEAE-Sepharose*

(i)   Equilibrate an 8 ml column of DEAE-Sepharose with Buffer A.
(ii)   Prepare the dextran-sulphate eluate for chromatography by dialysis against Buffer A.
(iii) Apply the dialysed extract to the column, wash and elute with a gradient of NaCl (from 0 to 0.5 M) in the same buffer.

Table 4. Purification of connective tissue metalloproteinases. The three metalloproteinases gelatinase, stromelysin and collagenase were purified from human gingival fibroblast culture medium according to the scheme shown in *Figure 5*. The activities eluting from each matrix were assayed against ^{14}C-labelled gelatin (77), ^{14}C-labelled casein (75) and ^{14}C-labelled collagen (74), active fractions pooled and analysis of the total number of units of activity and total protein [usually by A_{280} measurements, assuming an $E_{280}^{1\%}$ of 1, but alternatively by the Bradford Coomassie Blue binding method (87) for coloured preparations at the start of the protocol assessed] to give a value for the specific activity. The percentage recovery at each stage is also given. One unit of activity is 1 μg of the respective substrates degraded min^{-1}. n.a. = not applicable.

	Gelatinase		Stromelysin		Collagenase	
	Specific activity (units mg^{-1})	Recovery (%)	Specific activity (units mg^{-1})	Recovery (%)	Specific activity (units mg^{-1})	Recovery (%)
Culture medium	11.7	(100)	36	(100)	4.5	(100)
Dextran-sulphate	47.1	70	77	49	39.7	99
DEAE Sepharose	657	25	125	50	296	70
Heparin Sepharose	1233	18	88	18	2715	46
Gelatin Sepharose	6370	24	n.a.	−	n.a.	−
Red Sepharose	n.a.	−	322	16	n.a.	−
Zn^{2+} imino-diacetic acid Sepharose	n.a.	−	n.a.	−	4194	39
Ultrogel	12 410	21	366	17	8004	39

(iv) Collect fractions and assay for gelatinase using ^{14}C-labelled gelatin. Collagenase and stromelysin do not bind to this column and are thus present in the flow through. Keep this for further processing.

5.3.2 *Heparin Sepharose*

(i) Pool the gelatinase-containing fractions against Buffer A.
(ii) Equilibrate a column (1.5 ml) of heparin Sepharose with Buffer A.
(iii) Load the gelatinase-containing extract, wash and elute with a gradient of NaCl (from 0 to 1.0 M) in the same buffer.

It is interesting to note that gelatinase binds to this cationic matrix at the same pH as that used for the previous column.

5.3.4 *Gelatin Sepharose*

(i) Equilibrate a column (1.5 ml) of gelatin Sepharose with Buffer A containing 1 M NaCl.
(ii) Apply the pooled eluates from the previous step directly, wash.
(iii) Elute with a gradient of dimethyl sulphoxide (0−10%) in the equilibration buffer.

5.3.3 *Ultrogel AcA34 gel filtration*

(i) Equilibrate a 150 ml column (about 60 cm in length) with Buffer A containing 0.5 M NaCl.
(ii) Calibrate using marker proteins (e.g. yeast alcohol dehydrogenase, M_r 150 000; bovine serum albumin, M_r 66 000; ovalbumin, M_r 45 000, and carbonic anhydrase, M_r 29 000).

145

(iii)   Apply the pooled gelatinase-containing fractions from the previous step, preferably less than 5% of bed volume and elute.

This step is included in order to remove autodegradation fragments generated during previous steps.

5.4 Separation of stromelysin

5.4.1 *Heparin Sepharose*

(i)   Equilibrate a column (10 ml) of heparin Sepharose with Buffer A.
(ii)   Load the flow-through fraction from the DEAE Sepharose chromatography described in Section 5.3.1 directly on to this column.
(iii)   Wash and elute with a gradient of NaCl (0−0.5 M) in Buffer A.
(iv)   Assay fractions for stromelysin using ^{14}C-labelled casein and for collagenase using ^{14}C-labelled type I collagen.

The majority of the stromelysin does not bind to this column, whereas collagenase does. However, there is some variation with different batches of heparin Sepharose and therefore careful monitoring is necessary.

5.4.2 *Reactive Red Sepharose*

(i)   Equilibrate Red Sepharose (1.5 ml) with Buffer A.
(ii)   Apply the flow-through from the previous column and wash.
(iii)   Elute with a gradient of NaCl (0−0.5 M) in Buffer A.

Truncated forms of this enzyme which lack the C-terminal portion do not bind to this matrix (72,75). Active forms of this enzyme are eluted prior to the latent form and separation can be optimized by manipulation of the gradient conditions.

5.4.3 *Ultrogel AcA44 gel filtration*

The stromelysin produced by the above steps is frequently contaminated with low-M_r material which can be removed by gel filtration, essentially as described in Section 5.3.4.

5.5 Separation of collagenase

5.5.1 *Zn^{2+}-iminodiacetic acid Sepharose*

(i)   Equilibrate a column (1.5 ml) of Zn^{2+}-iminodiacetic acid Sepharose with Buffer A containing 0.5 M NaCl.
(ii)   Apply the collagenase-containing fractions from heparin Sepharose (Section 5.3.2) directly.
(iii)   Wash with Buffer A containing 0.5 M glycine or with Buffer B. (Buffer B = 25 mM sodium cacodylate pH 6.5 containing 0.5 M NaCl, 10 mM $CaCl_2$, 0.05% Brij 35 and 0.02% azide).
(iv)   Elute with 1 M glycine (in Buffer A or Buffer B). Alternatively, elution can be effected either with a gradient of imidazole (0 to 50 mM) in Buffer A or with a lower pH step elution (e.g. 50 mM sodium acetate pH 4.8 containing 0.5 M NaCl, 10 mM $CaCl_2$, 0.05% Brij 35 and 0.02% azide).

This matrix can be used instead of the Reactive Red Sepharose described in Section

5.4.2 for the purification of stromelysin (78) which is present in the wash (0.5 M glycine on Buffer B).

Stromelysin and collagenase are so similar in size and properties that it is difficult to separate them totally. Their elution profiles on heparin Sepharose, Zn^{2+}-iminodiacetic acid Sepharose and gel filtration matrices overlap and trace cross-contamination is difficult to avoid. If available, absorbing out with anti-collagenase IgG Sepharose (72) or anti-stromelysin IgG Sepharose is recommended.

Recently a new matrix comprising hydroxamic acid (81) has been described for the purification of collagenase. However, its affinity for the other metalloproteinases has yet to be clarified.

5.5.2 *Ultrogel AcA44 gel filtration*

As for stromelysin and gelatinase, the collagenase preparation is separated from low M_r fragments by filtration under the same conditions.

6. MONOCLONAL ANTIBODIES AND THEIR FRAGMENTS—M.Perry and H.Kirby

The aim of this section is to review, by example, the methods currently available for the purification of monoclonal antibodies and their immunoreactive fragments derived by enzyme cleavage.

The purification properties of monoclonal antibodies frequently differ considerably from those predicted by a knowledge of the species polyclonal. While the basic methodology of purification is comparable for both polyclonal and monoclonal antibodies, the conditions required to maintain maximum antibody activity are far stricter for the latter. The immunoglobulins present in the serum of an immunized animal originate from several distinct B-cell clones and are thus heterogeneous with respect to structure. Monoclonal antibodies, however, originate from a single B-cell clone and, with the exception of microheterogeneity due to variable glycosylation, may be considered to possess an identical structure. Thus, for a polyclonal antiserum the loss or inactivation, during purification, of antibodies from a single B-cell clone would largely be inapparent, whereas the same conditions would be disastrous for a given monoclonal antibody. Solubility and stability are often more restricted for monoclonal antibodies, with precipitation occurring at the pI and inactivation during elution at low pH. Particular attention has to be paid to the buffers employed in terms of composition, pH and ionic strength. Precipitation is commonly used in the isolation of polyclonal antibodies, but this method frequently results in irreversible damage to monoclonal antibodies.

Purification of monoclonal and polyclonal antibodies also differs in the nature of the starting material, and hence contaminating compounds, from which they require to be isolated. Monoclonal antibodies are obtained from hybridomas either by *in-vivo* propagation as ascitic tumours or by *in-vitro* tissue culture. The former method produces a much higher concentration of specific antibody, typically 5 mg ml^{-1} in ascitic fluid compared to 50 μg ml^{-1} in tissue culture supernatant. However, ascitic fluid also contains a significant amount of host immunoglobulin (up to 100% w/w of the specific immunoglobulin), whereas the specific monoclonal is the only hybridoma species immunoglobulin present in the tissue culture supernatant.

The culture medium usually contains a serum additive (typically 10% v/v foetal calf serum) which is the main source of contaminant protein in the tissue culture supernatants. The most troublesome contaminants are frequently the host immunoglobulins in ascitic fluids and the serum immunoglobulins (usually bovine) in tissue culture supernatant. However, the latter may soon be a problem of the past, due to progress in the use of serum-free culture media (88) in hybridoma technology.

Three criteria govern the choice of purification regime for monoclonal antibodies; they are, in order of priority:

(i)   purity of antibody required;
(ii)   volume of extract from which antibody is to be isolated;
(iii) type of extract from which antibody is to be isolated (e.g. ascites or tissue culture supernatant).

Thus a regime may be required to rapidly process a few millilitres of ascites to an intermediate level of antibody purity, whereas at the other extreme it may be required to isolate an antibody from many litres of tissue culture supernatant to 100% purity.

This section reviews the methods most suited to satisfying these criteria.

6.1 Extract clarification

Wherever possible it is recommended that cells and cellular debris be removed from ascitic fluid and tissue culture supernatant prior to commencing the purification regime. Ascitic fluid is centrifuged twice, the first time at 1000 g for 10 min to remove cells and then again at 50 000 g for 45 min to sediment debris. Where the volumes of tissue culture supernatant render this procedure impracticable, the extract can be processed by sequential filtration through 5 μm, 1 μm and finally 0.45 μm filters.

6.2 Selective precipitation

The selective precipitation of polyclonal antibodies from serum by salting-out procedures is often quoted as a simple, cheap and effective method of purification. The precipitants most commonly employed include sodium sulphate (20% w/v), ammonium sulphate (26% w/v), ethanol (50% v/v) and polyethylene glycol 6000 (13% w/v) (89,90). However, our experience with monoclonal antibodies, in agreement with Phillips *et al.* (91), is that, despite careful attention to pH, temperature and protein concentration, these methods cause significant damage with concomitant loss in activity. Our preferred route is to precipitate the contaminating proteins leaving the monoclonal in solution, using the following method.

(i)   Add ascitic fluid (x ml) to 50 mM acetate buffer pH 4.0 ($2x$ ml).
(ii)   Adjust the pH to 4.5 with 2 M HCl or NaOH.
(iii) Add caprylic acid ($x/15$ g) slowly with constant stirring.
(iv)   Continue stirring for a further 30 min.
(v)   Sediment the precipitate by centrifugation at 1000 g for 10 min.
(vi)   Remove the supernatant and adjust the pH to 6.0 with 2 M NaOH.
(vii) Dialyse against a buffer appropriate for the monoclonal antibody.

Advantages. This is a rapid method of purifying antibodies of IgG sub-class to approximately 80−90% purity with negligible loss of activity.

Disadvantages. The method is not suited to tissue culture supernatant, because of its low yields and high final dilution factor. This can be obviated to a degree by prior concentration of the tissue culture supernatants (requires a minimum final antibody concentration of 500 μg ml^{-1}).

6.3 Affinity chromatography

Affinity chromatography involves the specific interaction between an insoluble ligand and the monoclonal antibody which can be subsequently disrupted to yield pure monoclonal antibody. Three types of ligand are commonly employed:

(i)   Protein A or Protein G;
(ii)   antibody directed against the monoclonal species (usually anti-mouse IgG);
(iii)   antigen to which the monoclonal antibody was raised.

These are discussed below.

6.3.1 *Protein A*

Protein A binds the F_c region of the major sub-classes of mouse and rat IgG with varying affinities (92,93). The binding of Protein A is enhanced by ionic strength, particularly phosphate. An example of a method is given below.

(i)   Slowly add 1 M phosphate buffer pH 8.0 (x ml) to the ascitic fluid or tissue culture supernatant (x ml).
(ii)   Apply the solution to a column of Protein A Sepharose ($x/10$ ml for tissue culture supernatant; $10x$ ml for ascitic fluid) equilibrated with 0.5 M phosphate buffer pH 8.0.
(iii)   Wash the column with two bed volumes of 0.5 M phosphate buffer pH 8.0 or until the A_{280} of the column eluent has returned to zero.
(iv)   Commence collecting fractions ($x/20$ ml for tissue culture supernatants, x ml for ascites).
(v)   Elute with two bed volumes of 0.1 M citrate buffer, pH 6.0.
(vi)   Elute with two bed volumes of 0.1 M citrate buffer, pH 4.5.
(vii)   Elute with two bed volumes of 0.1 M citrate buffer, pH 3.5.
(viii) Add 1/4 volume of 1 M Tris−HCl, pH 8.0, to each fraction.
(ix)   Pool those fractions containing the antibody and adjust the pH to 7.5 with 2 M NaOH.
(x)   Exchange (e.g. dialyse) the antibody into the desired buffer.

Typically, mouse monoclonals IgM, IgA and IgG$_3$ do not bind to the column. IgG$_1$ is eluted at pH 6.0, IgG$_{2a}$ at pH 4.5 and IgG$_{2b}$ at pH 3.5. However, in agreement with the results of Stephenson *et al.* (94), we have noted (i) certain monoclonals eluting at more than one pH (e.g. 70% at pH 4.5 and 30% at pH 3.5), and (ii) atypical elution pH values (e.g. IgG$_{2a}$ requiring pH 3.5 for elution).

Advantages. This is a rapid one-step purification which removes most non-IgG contaminants and can achieve purities greater than 99%. It is particularly useful for large-scale purifications of tissue culture supernatant, where 10- to 100-fold concentration can be achieved. Elution conditions (pH 6.0 and pH 4.5) are relatively mild, although some monoclonals may be damaged at pH 3.5.

Disadvantages. When used to purify ascitic fluid, host immunoglobulins are also bound by Protein A. Resolution of specific monoclonal from the host IgG can sometimes be achieved by use of a pH gradient rather than the stepwise elution detailed above. Quality of Protein A Sepharose (or equivalent) is important to avoid leakage of Protein A during the elution procedure. In our experience, those immobilized Protein A resins most resistant to leakage are those in which Protein A is linked via an amide bond to the matrix. Examples of such resins are Eupergit-Protein A (Rohm Pharma) and Affi-Prep Protein A (Bio-Rad) in which the linkage is formed as a result of the reaction between the amino groups of Protein A and either oxirane or N-hydroxysuccinimide ester groups respectively. Certain monoclonals aggregate or are inactivated by high phosphate concentrations. This can be circumvented by the replacement of the 1 M phosphate buffer with 20 mM Tris−HCl, pH 8.0, containing 1 M NaCl and replacement of the 0.5 M phosphate with 20 mM Tris−HCl, pH 8.0, containing 0.5 M NaCl.

6.3.2 *Protein G*

Protein G is a protein originally isolated from the cell walls of human group G streptococci, and is analogous to Protein A in that it binds strongly to the F_c region of IgG (95). It is available immobilized to agarose (Perstorp Biolytica). In comparison with Protein A, it offers two principal advantages:

(i)   It is applicable to a wider range of immunoglobulins, in particular Rat IgG_{2A}, IgG_{2B}, Human IgG_3, Bovine IgG_1 and sheep IgG_1.

(ii) It requires a starting buffer of lower ionic strength, typically 10 mM phosphate pH 7.5 containing 0.15 M NaCl.

One potential disadvantage compared with Protein A is that, in our limited experience, a lower pH is required to elute a given monoclonal antibody from Protein G-agarose.

6.3.3 *Anti-mouse Ig*

Immobilized anti-mouse Ig (e.g. on Sepharose) is used to bind the monoclonal in a method analogous to Protein A. However, the affinity of the anti-mouse Ig is approximately two orders of magnitude greater than that of Protein A. Hence elution conditions for the former are much more rigorous (typically 0.1 M glycine−HCl, pH 2.5), and consequently damage to the monoclonal is more likely. One method of reducing this inactivation is to use low affinity anti-mouse Ig antibodies. Anti-mouse antibodies are readily available from a number of suppliers. It is preferred that the host be either sheep or goat and that the antiserum be raised to the F_c region of the mouse immunoglobulin). These can be prepared as follows:

(i)   Add 0.1 M phosphate buffer, pH 8.0 (*x* ml) to the anti-mouse Ig antibodies (*x* ml).

(ii)   Apply the solution to a column of mouse IgG-Sepharose (*x* ml^{-1}) equilibrated with 0.1 M phosphate buffer pH 8.0.

(iii) Wash the column with 0.1 M phosphate until the A_{280} value returns to zero.

(iv)   Commence collecting 1 ml fractions.

(v)   Elute with two bed volumes of 0.1 M citrate buffer, pH 3.5.

(vi)   Add 1/4 volume of 1 M Tris−HCl pH 8.0 to each fraction.

(vii) Pool those fractions containing the anti-mouse IgG antibodies eluted at pH 3.5.

(viii) Dialyse into 0.1 M NaHCO$_3$−Na$_2$CO$_3$, pH 9.1.

The low-affinity anti-mouse Ig antibodies thus obtained can then be coupled to Sepharose (or the equivalent) to give a suitable affinity column. Purification of the monoclonal is then as described for Protein A, with the exception that 0.1 M phosphate, pH 8.0, replaces the 1 M and 0.5 M phosphate buffers, and the column is eluted directly with 0.1 M citrate, pH 3.5, omitting the pH 6 and pH 4.5 citrate buffers.

Advantages. This method can be used to isolate mouse IgM and IgG$_3$ which frequently fail to bind to Protein A-Sepharose. It may be of use in isolating the monoclonal antibody from supernatants which contain high levels of serum Ig (for example, those from bovine tissue cultures).

Disadvantage. Preparation of low-affinity anti-mouse antibodies is time-consuming and yields are frequently low. The procedure is rarely cost-effective.

6.3.4 Specific antigen

The use of immobilized antigen to purify specific antibodies is well known (96). However, the conditions required to disrupt the antigen−antibody binding may be harsh (chaotrophic ions, organic acids with low surface tension, denaturants and extreme pH values), and in our experience is limited to monoclonals of low affinity, typically where $K_d = 10^{-8}$ M. However, when it is applicable it is capable of purifying a monoclonal to better than 99% purity in a single step, even in the presence of host Ig. Unlike the other methods, this method is capable of distinguishing monoclonals with intact binding activity from those which are denatured.

6.4 Ion-Exchange chromatography

Mouse monoclonal antibodies exhibit much heterogeneity in isoelectric point, ranging, in our experience, from pH 4.8 to 8.4 and having no correlation with isotype. With the advent of the high-performance resins such as Mono S and Mono Q (Pharmacia), ion-exchange chromatography provides an excellent method of purification, and one that can be rapidly tailored to suit the requirements of each individual monoclonal antibody.

   Classically, immunoglobulins have been purified from serum by anion-exchange chromatography (97). However, phenol red, a pH indicator commonly used in cell culture media, binds very strongly and partially irreversibly to anion-exchangers, reducing both the capacity of the resin and its lifetime. Similarly, albumin, the major contaminant in both ascites and tissue cultures, binds more strongly than the monoclonals, further reducing the capacity of the resin. Furthermore, it is necessary to dialyse the ascites or tissue culture supernatant against the low-ionic-strength starting buffer, a time-consuming procedure which can lead to precipitation and denaturation of the monoclonal. Therefore, because of their high capacity for monoclonal antibodies relative to the major impurities, we find that cation-exchangers are far better suited to the isolation of monoclonal antibodies from both ascitic fluid and tissue culture supernatants. A typical procedure is given below.

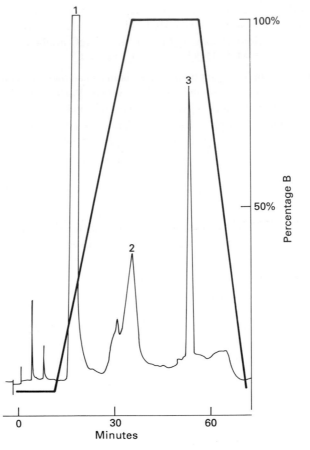

Figure 6. S-Sepharose separation of monoclonal antibody from tissue culture supernatant.

Column:	S-Sepharose (4 × 1.8 cm)
Sample:	Tissue culture supernatant containing monoclonal antibody BCD-1 (× 10 concentrate)
Buffers:	(**A**) 50 mM Mes, pH 6.0 (**B**) 50 mM Mes, 0.25 M NaCl, pH 6.0
Flow:	2 ml min^{-1}
Detection:	UV at 280 nm
Peak identification:	(**1**) Albumin (**2**) Transferrin (**3**) BCD-1

6.4.1 *Cation-exchange chromatography*

(i)   Add distilled water (*x* ml) containing 0.02% Tween 80 to the ascitic fluid (*x* ml) or tissue culture supernatant (*x* ml). Adjust the pH to 6.0 with 2 M HCl.

(ii)   Apply to a column of S-Sepharose (3*x* ml for ascitic fluid, *x*/20 ml for tissue culture supernatant) equilibrated with 50 mM Mes, pH 6.0.

152

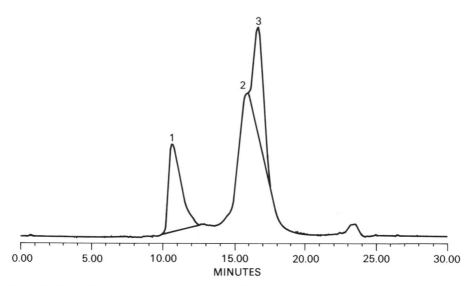

Figure 7. Use of HPLC gel-filtration to monitor antibody aggregation.

Column:	TSK 3000 SW (30 × 0.8 cm)
Sample:	Monoclonal antibody BCD-3 after overnight incubation in 1 M phosphate pH 8.0.
Buffers:	0.25 M Mes pH 6.0
Flow:	0.5 ml min^{-1}
Detection:	UV at 280 nm
Peak identification:	(**1**) BCD-3 aggregate (**2**) BCD-3 dimer (**3**) BCD-3

(iii)   Wash the column with 50 mM Mes, pH 6.0, until the A_{280} value of the eluent has returned to zero.

(iv)   Commence collecting fractions of the appropriate volume, typically 1/10 of the bed volume.

(v)   Elute with a linear gradient (10 column volumes) from 0 to 0.5 M NaCl, in 50 mM Mes, pH 6. On completion of the gradient, continue elution with 50 mM Mes, pH 6.0, containing 0.5 M NaCl, for a further 2 bed volumes.

(vi)   Pool those fractions containing the monoclonal antibody, and exchange (e.g. dialyse) into the appropriate storage buffer.

A typical elution profile is shown in *Figure 6*. As can be seen, the major contaminants, including albumin, are eluted early, well before the monoclonal antibody. In our experience, transferrin is the most likely contaminant to co-elute with the monoclonal, and thus the shape and slope of the elution gradient must be optimized for each individual monoclonal. We find this is most easily accomplished by establishing the gradient conditions using analytical HPLC on Mono S columns (Pharmacia) which exhibit similar ion-exchange characteristics to S-Sepharose.

153

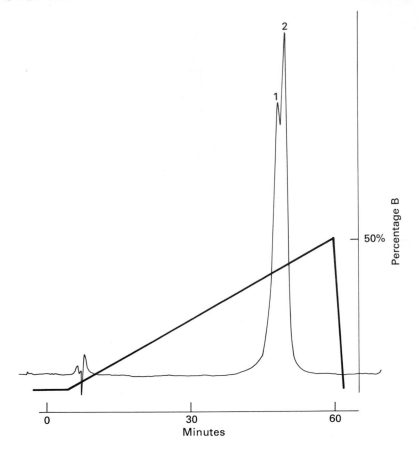

Figure 8. Mono S resolution of monoclonal antibody microheterogeneity.

Column:	Mono-S HR 5/5
Sample:	Monoclonal antibody BCD-2 purified by protein A
Buffers:	(**A**) 50 mM Mes pH 6.0 (**B**) 50 mM Mes, 0.25 M NaCl, pH 6.0
Flow:	1 ml min^{-1}
Detection:	UV at 280 nm
Peak identification:	(**1**) BCD-2 (**2**) BCD-2

The only detrimental effect likely to be suffered by the monoclonal is as a result of aggregation. For this reason, we routinely screen the purified monoclonals by HPLC gel filtration, a rapid and convenient method of monitoring aggregation (*Figure 7*). Aggregation is usually a result of chromatography performed at a pH too close to the isoelectric point of the monoclonal. Although pH 6.0 is a useful starting point, the

154

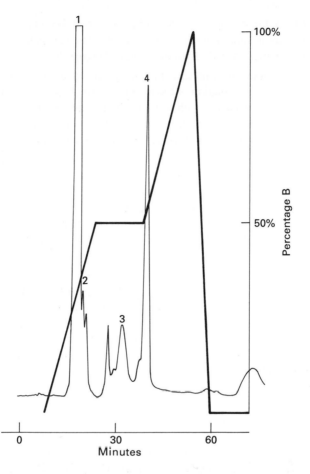

Figure 9. S-Sepharose separation of monoclonal antibody from host immunoglobulin.

Column:	S-Sepharose (4 × 1.8 cm)
Sample:	Mouse ascites containing monoclonal antibody BCD-1
Buffers:	(A) 50 mM Mes pH 6.0
	(B) 50 mM Mes, 0.25 M NaCl, pH 6.0
Flow:	2 ml min^{-1}
Detection:	UV at 280 nm
Peak identification:	(1) Albumin
	(2) Transferrin
	(3) Mouse immunoglobulin
	(4) BCD-1

chromatography can be performed over the range pH 4.0−8.5 using the following buffers; 50 mM acetate (pH 4.0−5.5), 50 mM Mes (pH 5.5−7.0), 50 mM Hepes (pH 7.0−8.5).

We have observed that, due to the resolving power of S-Sepharose and, especially, Mono S, certain monoclonals are resolved into more than one peak, presumably because of microheterogeneities in carbohydrate content (*Figure 8*).

Advantages. In our hands, we find cation-exchange chromatography superior to Protein A chromatography for the isolation of monoclonals from both ascites and tissue culture. It frequently gives better yields, with less inactivation of the antibody. It is applicable to all Ig classes and IgG isotypes and, in the case of ascites, can resolve the monoclonal from the host immunoglobulins (*Figure 9*). When used to purify tissue culture supernatant, the ability of this technique to deliver the monoclonal in a higly concentrated form is of particular value.

Disadvantages. In order to obtain the highest purity, the chromatography requires 'fine-tuning' for each individual monoclonal antibody. Dilution is required prior to chromatography in order to lower the ionic strength of the sample. This may prove a problem for large-scale purifications.

6.4.2 *Mixed-mode chromatography*

Another form of ion-exchange chromatography particularly applicable to monoclonal antibodies is mixed-mode chromatography. Mixed-mode ion-exchangers contain both weakly anionic and weakly cationic functional groups. Examples of such resins are Bakerbond ABX (J.T.Baker) and hydroxylapatite (Bio-Rad).

Protein adsorption to hydroxylapatite, a form of calcium phosphate, is largely due to interaction between negatively charged protein moieties and Ca^{2+} groups, whereas interaction between positively charged protein groups and phosphate groups is of less importance (98). Generally, as with cation ion-exchange, monoclonal antibodies are more strongly bound to hydroxylapatite than are contaminating proteins such as albumin and transferrin. Thus the binding capacity is high for the antibody, which is usually adsorbed to the resin in a low-ionic-strength phosphate buffer at neutral pH and subsequently eluted by increasing ionic strength, typically to 200 mM phosphate (99).

In most respects ABX performs in a manner similar to hydroxylapatite, but does have advantages in terms of higher capacity (150 mg IgG g^{-1} ABX), increased mechanical strength, increased column lifetime and cost.

Advantages. These are the same as for cation-exchange chromatography.

Disadvantages. These are the same as for cation-exchange chromatography.

6.5 Hydrophobic interaction chromatography (HIC)

The hydration of salt ions, such as ammonium sulphate, requires a considerable part of the solvent, lowering the solvation of proteins and exposing non-polar amino acid side chains on the protein. These exposed side chains interact with hydrophobic groups of the resin, typically polyethylene glycol (Hydropore-HIC; Rainin) or phenyl groups (Bio-Gel TSK Phenyl-5-PW; Bio-Rad) (100). Thus the antibodies are bound to the column in high salt concentration (typically 3.0 M ammonium sulphate in 0.1 M phosphate, pH 7.0) and subsequently eluted in order of increasing hydrophobicity with a gradient of decreasing salt concentration (limit buffer typically 0.1 M phosphate, pH 7.0).

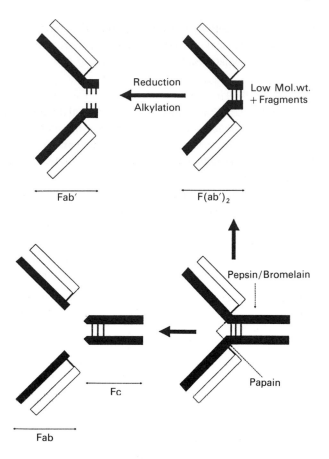

Figure 10. Antibody fragments from mouse IgG. Pepsin and bromelain cleave the heavy chains in the C-terminus region of the hinge to yield F(ab')$_2$ and low-molecular-weight peptides or pF$_c$' fragments. Reductive alkylation of the F(ab')$_2$ yields two Fab' fragments. Papain splits the molecule into two Fab fragments and an F$_c$ fragment in the hinge region.

Certain weak ion-exchangers such as Hydropore-AX, a weak anion exchanger, and Hydropore-SCX, a weak cation exchanger (both by Rainin) can also be used for hydrophobic chromatography. At high salt conditions (typically 3.0 M ammonium sulphate), charge interactions between these resins are eliminated and the supports function as hydrophobic matrices. This 'multi-model' use of a single resin can be of particular use in achieving very high degrees of purity. For example, the column can be used as an ion-exchange resin to achieve a first-stage purification. Ammonium sulphate is then added to the pooled fraction which is then rechromatographed on the same column, but in the HIC mode to achieve the final purification (see manufacturer's literature for details).

Advantages. This technique can prove useful when used in combination with ion-exchange chromatography.

Disadvantages. Precipitation of monoclonal antibodies can occur at ammonium sulphate concentrations required to permit hydrophobic binding to the matrix. HIC requires a degree of denaturation of the antibody to permit exposure of hydrophobic side chains. In general, the loss of monoclonal antibody activity is greater with HIC than with ion-exchange or Protein A chromatography.

6.6 **Fragment preparation**

The immunoreactive fragments F(ab')$_2$ and Fab' are separated from the native antibody by controlled proteolytic cleavage at the hinge region of the molecule (101,103). The divalent F(ab')$_2$ can be further separated into univalent Fab' fragments by chemical reduction of the hinge disulphide bonds (102) (*Figure 10*). Conditions for proteolytic digestion of antibody are well established. Although pepsin is the most commonly used enzyme for F(ab')$_2$ preparation, some antibodies are completely degraded by it, and in our experience by cysteine proteases papain and bromelain are often more effective. The resulting F(ab')$_2$ fragments, irrespective of antibody sub-class, vary considerably in their properties, depending on the specificity of the enzyme used (*Table 5*). Factors such as immunoreactivity, stability and charge are affected, and in the light of this variability it is advisable to establish methods for preparation and purification individually for each antibody.

Preliminary evaluation of the digest is usually done on an analytical scale (1 mg antibody). Reaction conditions are optimized with respect to the rate of digestion and retention of immunoreactivity, the course of the reaction being most conveniently monitored by analytical gel filtration HPLC on a Zorbax GF-250 column. Typical reaction protocols for F(ab')$_2$ preparation are described below.

6.6.1 *Pepsin digestion*

(i)   For mouse monoclonal antibodies, use a pH range 3−4.8 in 0.1 M sodium acetate buffer.

(ii)   Vary the reaction ratios of antibody to enzyme by weight (e.g. 20:1, 10:1, 5:1).

(iii)   React at 37°C in a water-bath. Remove 25 μl aliquots to monitor the reaction at time intervals.

(iv)   Stop the reaction by addition of 1 M Tris base to raise the pH to about 8.0.

The highest grade pepsin from several suppliers was assessed and no significant differences were observed.

Parameters to be varied to optimize the reaction are pH, reaction time, temperature and ratio of reactants. Pepsin autodigests, therefore fresh enzyme will have to be added after an overnight incubation. Protein precipitation is often seen at concentrations above 5 mg ml^{-1}. The addition of 0.25 M NaCl prevents this (*Table 5*).

6.6.2 *Bromelain digestion*

(i)   Bromelain activation: react 2.5 ml of enzyme (about 5 mg) in 0.1 M sodium acetate pH 5.5 with 50 mM cysteine at 37°C for 30 min. Desalt (e.g. using Sephadex, PD-10) into 0.1 M acetate buffer pH 5.5, containing 3 mM EDTA.

(ii)   Vary the reaction ratios of activated enzyme to antibody as described above with pepsin.
React at 37°C and monitor by HPLC gel filtration.

Table 5. F(ab')$_2$ preparations with bromelain or pepsin.

Pepsin and/or bromelain were used to produce F(ab')$_2$ fragments on a range of mouse MAbs. Products of proteolytic digestion were analysed by gel-permeation hplc (gpc) for size, and ion-exchange chromatography (iec) for charge. Stability and immunoreactivity were assessed using appropriate immunoassays.

MAb	Sub-class	Enzyme	Charge (iec)	Size (gpc)	Stability	Comment
1	IgG-1	1. Bromelain	Single species	Single peak	Good	100% retention of activity.
		2. Pepsin	–	–	–	Completely degraded
2	IgG-1	1. Bromelain	Single species	Single peak	Good	No loss in activity
		2. Pepsin	2 distinct species	Single peak	Variable	Early eluting peak on iec less stable than second peak
3	IgG-2a	1. Pepsin	Single species	Single peak	Good	No loss in activity
4	Not determined	1. Pepsin	Single species	Single peak	Good	Scale up of MAb > 5 mg ml^{-1} required 0.25 M NaCl to prevent precipitation. No loss in activity
5	IgG-2a	Pepsin	Single species	Single peak	Good	Slight loss in activity
6	IgG-2a	Pepsin	Single species	Single peak	Good	No loss in activity
7	IgG-3	Pepsin	–	–	–	No identifiable product Ab completely degraded
8	IgG-1	Pepsin	2 distinct species	Single peak	Not assessed	70% activity in total F(ab')$_2$ product
9	IgG-2a	Pepsin	Single species	Single peak	Good	No loss in activity
10	IgG-1	1. Pepsin	Single species	Single peak	Good	No loss in activity
		2. Bromelain	Single species	Single peak	Good	No loss in activity
11	IgG-1	1. Pepsin	–	–	–	Totally degraded
		2. Bromelain	Single species	Single peak	Good	No identifiable product
12	IgG-1	1. Pepsin	–	–	–	Totally degraded
		2. Bromelain	Single species	Single peak	Good	No loss in activity

(iv)   Stop the reaction by adding protease inhibitors leupeptin and antipain to a final concentration of 50 μg ml^{-1}.

Fab' is readily produced from F(ab')$_2$ by mild reductive alkylation (103). Bromelain is available commercially either as a freeze-dried powder or in suspension. The quality of the enzyme varies with supplier. In our experience, bromelain from BCL is the most active, although some batch variation has been noticed.

6.7 Purification of fragments

A typical proteolytic digest of an antibody consists of several components: antibody fragments, native antibody, native or hydrolysed enzyme and low molecular-weight peptides, as well as inhibitors and alkylating agents. Separation of the required antibody fragment from the other contaminants is readily achieved by conventional column chromatography on soft gels, or by HPLC/FPLC. Ultrafiltration is also appropriate under some circumstances. As with intact antibody purification, care should be taken with choice of buffer composition so that problems such as fragment aggregation or precipitation can be minimized. The main factors which govern the method of purification are:

(i)   the amount of material to be purified;
(ii)   the degree of purity required;
(iii) yield.

One of the principal limitations constantly encountered when working with monoclonal antibodies and their fragments is the small quantity of starting material available (10 mg). In our experience, for most laboratory-scale fragment purifications HPLC/FPLC methods are appropriate, and these are discussed in this section, with suggestions for scale-up included where relevant.

6.7.1 *Gel filtration*

This method is suitable for separating Fab' from F(ab')$_2$ fragments (M_r 40 000 and 100 000 respectively) where baseline resolution is achieved by gel-filtration hplc on TSK G3000 SW or Zorbax GF-250-XL (21 mm $\times$ 30 cm), using the following method.

(i)   Equilibrate column in 0.1 M sodium phosphate pH 7.0 at a flow rate of 2 ml min^{-1}.
(ii)   Filter sample (2 $-$ 50 mg protein) using a 0.2 μm membrane filter.
(iii) Apply to column in a maximum sample volume of 2 ml and elute at a flow rate of 2 ml min^{-1}.
(iv)   Collect appropriate fractions and concentrate to required level using an Amicon stirred cell or Centricon CM10, depending on the volume to be concentrated.
(v)   Analyse the product for purity by re-inspection on an analytical gel filtration column (Zorbax GF-250) and by SDS-PAGE. Typical yields obtained are 90%. For larger-scale purification, Sephacryl S200 (Pharmacia), with high resolution and fast flow properties, or Ultrogel AcA-44, may be used in a column of suitable size.

F(ab')$_2$ can also be purified by gel filtration, but a one-step purification of this type is really only successful if the reaction has been reliably optimized such that no intact

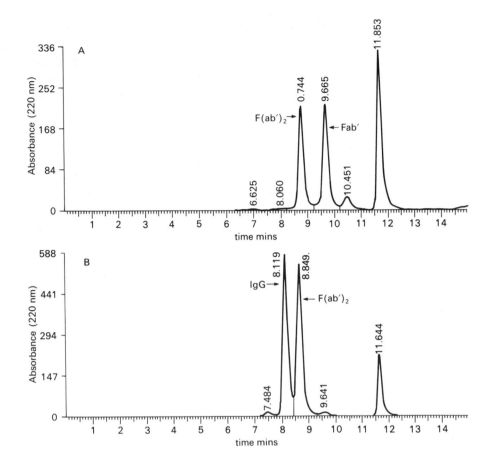

Figure 11. HPLC filtration profiles of native antibody and F(ab')$_2$ fragment. (**A**) Resolution between F(ab')$_2$ (M$_r$ 100 000) and Fab' (M$_r$ 50 000) showing good baseline discrimination. (**B**) Resolution between intact IgG and F(ab')$_2$ fragment, showing poor baseline separation.

Column:	Zorbax GF-250
Flow rate:	1 ml min^{-1}
Buffer:	0.2 M sodium phosphate, pH 7.0

antibody remains, as resolution between intact antibody and F(ab')$_2$ is poor (*Figure 11*). In our experience, it is only rarely possible to guarantee total absence of intact antibody, and if degree of purity is an important factor, then one of the following methods is more suitable.

6.7.2 *Affinity chromatography*

The affinity ligand most conveniently employed to purify F(ab')$_2$ from a proteolytic digest of antibody is Protein A Sepharose. This is essentially a passive purification where F(ab')$_2$ fragments do not bind to the ligand and F$_c$ fragments or intact antibody do.

6.7.3 *Protein A Sepharose and gel filtration*

When, as is often the case, some intact undigested antibody remains, a Protein A adsorption stage to remove this contaminating antibody, followed by gel filtration to separate F(ab')$_2$ from enzyme, can be used as described below.

(i)   Equilibrate approximately 1 ml of gel with 0.1 M Tris−HCl, pH 8.0 (10 × volume buffer to gel) with several changes of buffer. Decant excess buffer.

(ii)   Using Protein A at 1 ml gel volume per 1 mg of original antibody, add the digest to the gel in a suitable vessel (e.g. centrifuge tube) and agitate gently for 30 min at room temperature, ensuring the gel is in suspension all the time.

(iii)   Centrifuge the reaction mixture at 600 g for 5 min.

(iv)   Carefully pipette out the supernatant and load onto one of the liquid chromatography columns recommended above.

Protein A used in this way minimizes any dilution effects of the sample, and excellent purity of F(ab')$_2$ is obtained in good yield. Alternatively, a small column can be used. Protein A-Sepharose alone is suitable only if there is no initial enzyme remaining, as seen with some pepsin digests. Pepsin auto-degrades into small peptide fragments. F(ab')$_2$ purified from such a reaction will contain small peptides, and a one-step purification of this type would be suitable only if this is not a concern.

It is worth noting that some mouse Ig sub-classes do not bind to Protein A, and also that F(ab')$_2$ from two mouse monoclonal antibodies has been reported to bind to Protein A (104).

6.7.4 *Ion-exchange chromatography*

As with intact monoclonal antibodies, F(ab')$_2$ fragments display a range of isoelectric points which vary with the type of enzyme used in the hydrolysis as well as with antibody sub-class, so that the choice of ion-exchange resin depends on the fragment in question. In our experience, anion exchange chromatography (DEAE, Mono Q) has been suitable for purifying most F(ab')$_2$ fragments from mouse monoclonal antibodies in a one-step purification at high purity and yield.

Because of the high capacity of ion-exchange resins, preliminary analytical work as well as laboratory-scale (∼30 mg antibody) batches are easily handled on a TSK DEAE-5PW column (8 mm × 7.5 cm). Although conditions should be optimized for each individual antibody, the following is a typical protocol.

(i)   Equilibrate the column in 10 column volumes of 20 mM Tris−HCl, pH 7.7 (Buffer A) at a flow rate of 1 ml min^{-1}.

(ii)   Desalt the antibody−enzyme digest into Buffer A on a small column of Sephadex G-25 or by dialysis. Filter through 0.5 μm filter.

(iii)   Load the sample onto the column at 1 ml min^{-1} and elute with Buffer A containing 0.25 M NaCl (Buffer B) under the following gradient conditions.

Time	% Buffer B
0−2 min	0
2−20 min	100
20−25 min	100
25−30 min	0

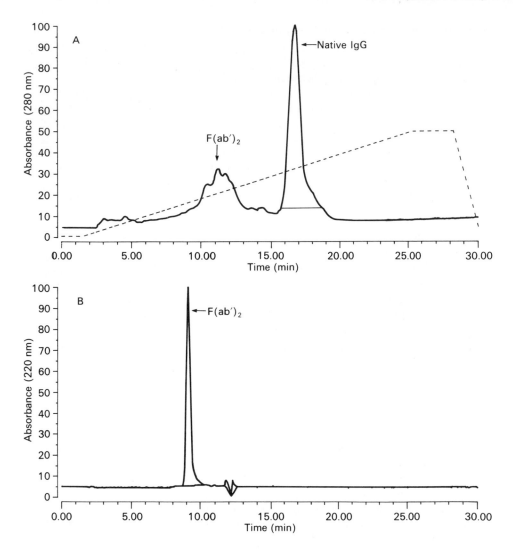

Figure 12. HPLC elution profiles of native antibody and corresponding F(ab')$_2$ fragment after pepsin digestion. (**A**) F(ab')$_2$ fragment and native antibody are separated by ion-exchange chromatography using a TSK DEAE -5PW column. The F(ab')$_2$ fragment is heterogeneous and shows multiple peaks (elution buffer, 0.5 M NaCl in 20 mM Tris−HC1, pH 7.7). (**B**) The F(ab')$_2$ peak purified from above elutes as a single sharp peak by hplc gel filtration on a Zorbax GF-250 column.

A typical elution profile is shown in *Figure 12*. The main contaminant with a pepsin digest is undigested antibody which is readily separated from F(ab')$_2$ fragments using this method. F(ab')$_2$ from pepsin digests tends to be fairly heterogeneous in terms of charge, and thus elutes as a single broad peak or multiple small peaks by ion-exchange, despite being a single sharp peak on gel filtration. This heterogeneity is presumably due to multiple cleavage sites on the antibody. In our experience, two particular mouse monoclonals (sub-class IgG$_1$, IgG$_{2a}$) formed two distinct populations each of F(ab')$_2$

163

with pepsin; one which did not bind and eluted at a retention time of 2 min and a second which eluted at 12 min, although they were indistinguishable in terms of size by gel filtration and non-reducing SDS-PAGE. However, there were differences in immunoreactivity and stability between them (*Table 5*).

In contrast, treatment with bromelain produces a more uniform product in terms of size and charge (*Table 5*). A note of caution here is that bromelain does not bind to the anion-exchanger under these conditions and may not be separated from weak binding F(ab')$_2$. A preceding gel-filtration step may have been included in this case (bromelain, M_r 28 000).

This method is easily scaled up. Excellent large-scale purification of F(ab')$_2$ is achieved on the fast-flow soft gel Q-Sepharose (Pharmacia) or Mono Q at 90% yield using this one-step method of purification.

6.7.5 *Ultrafiltration*

With a pepsin digest, F$_c$ fragments and pepsin tend to be broken down into small peptide fragments of M_r <5000. Therefore, if the digest has gone to completion with no intact antibody remaining, F(ab')$_2$ fragments can be separated from pepsin and peptide fragments by ultrafiltration. This is particularly useful for small-scale preparations where dilution and yield by liquid chromatography are problems. The Centricon CM30 micro-concentrators by Amicon (30 000 mol. wt cut-off) have been used very successfully on sample volumes of 2 ml or less.

7. ACKNOWLEDGEMENTS

S.J. Froud wishes to thank Dr Rosemary Froud for useful discussions and data with respect to the mammalian cytochrome oxidases.

8. REFERENCES

1. Harris,E.L.V. and Angal,S. (eds) (1989) *Protein Purification Methods: A Practical Approach*. IRL Press, Oxford.
2. Cross,A.R. and Anthony,C. (1980) *Biochem. J.*, **192**, 421.
3. Froud,S.J. and Anthony,C. (1984) *J. Gen. Microbiol.*, **130**, 3319.
4. Beardmore-Gray,M., O'Keeffe,D.T. and Anthony,C. (1983) *J. Gen. Microbiol.*, **129**, 923.
5. Wood,P.M. (1980) *Biochem. J.*, **189**, 385.
6. Thomas,P.E., Ryan,D. and Levin,W. (1976) *Anal. Biochem.*, **75**, 168.
7. Beardmore-Gray,M., O'Keeffe,D.T. and Anthony,C. (1982) *Biochem. J.*, **207**, 161.
8. Froud,S.J. and Anthony,C. (1984) *J. Gen. Microbiol.*, **130**, 2201.
9. Penefsky,H.S. and Tzagoloff,A. (1971) In *Methods in Enzymology*. Jakoby,W.B. (ed.), Academic Press, London and New York, Vol. 22, p. 204.
10. deVrij,W., Azzi,A. and Konings,W.N. (1983) *Eur. J. Biochem.*, **131**, 97.
11. Saraste,M. (1983) *Trends Biochem. Sci.*, **8**, 139.
12. Froud,R.J. and Ragan,C.I. (1984) *Biochem. J.*, **217**, 561.
13. Hatefi,Y. and Rieske,J.S. (1967) In *Methods in Enzymology*. Estabrook,R.W. and Pullman,M.E. (eds), Academic Press, London and New York, Vol. 10, p. 235.
14. Rieske,J.S. (1967) In *Methods in Enzymology*. Estabrook,R.W. and Pullman,M.E. (eds), Academic Press, London and New York, Vol. 10, p. 239.
15. Capaldi,R.A. and Hayashi,H. (1972) *FEBS Lett.*, **26**, 261.
16. Laemmli,U.K. (1970) *Nature*, **227**, 680.
17. Cohen,S. (1962) *J. Biol. Chem.*, **237**, 1555.
18. Levi-Montalcini,R. and Hamburger,V. (1951) *J. Exp. Zool.*, **116**, 321.
19. Rosenthal,A., Lindquist,P.B., Bringman,T.S., Goeddel,D.V. and Derynck,R. (1986) *Cell*, **46**, 301.
20. Kaplan,P.L., Topp,W.C. and Ozanne,B. (1981) *Virology*, **108**, 484.

21. Kaplan,P.L. and Ozanne,B. (1982) *Virology,* **123**, 372.
22. Shing,Y., Folkman,J., Sullivan,R., Butterfield,J., Murray,J. and Klaggbrun,M. (1984) *Science,* **223**, 1296.
23. Fernandez-Pol,J.A. (1985) *J. Biol. Chem.,* **260**, 5003.
24. Baird,A., Böhlen,P., Ling,N. and Guillemin,R. (1985) *Regulatory Peptides,* **10**, 309.
25. Huang,J.S., Huang,S.S. and Deul,T.F. (1983) *J. Cell Biol.,* **97**, 383.
26. Towbin,H., Staehelin,T. and Gordon,J. (1979) *Proc. Natl. Acad. Sci. USA,* **76**, 4350.
27. McAuslan,B.F., Bender,V., Reilly,W. and Moss,B.A. (1985) *Cell Biology International Reports,* **9**, 175.
28. Savage,C.R.,Jr and Cohen,S. (1972) *J. Biol. Chem.,* **247**, 7609.
29. Gregory,H. and Willshire,I.R. (1975) *Hoppe-Seyler's Z. Physiol. Chem.,* **309**, 1765.
30. Massague,J. (1983) *J. Biol. Chem.,* **258**, 13606.
31. Schreiber,A.B., Winkler,M.E. and Derynck,R. (1986) *Science,* **232**, 1250.
32. Massague,J. (1983) *J. Biol. Chem.,* **258**, 13614.
33. Derynck,R., Roberts,A.B., Winkler,M.E., Chen,E.Y. and Goeddel,D.V. (1984) *Cell,* **38**, 287.
34. Roberts,A.B., Anzano,M.A., Meyers,C.A., Widerman,J., Blacher,R., Pan,C.E., Stein,S., Lehrman,S.R., Smith,J.M., Lamb,L.C. and Sporn,M.B. (1983) *Biochem.,* **22**, 5692.
35. Sporn,M.B., Roberts,A.B., Wakefield,L.M. and Assoian,R.K. (1986) *Science,* **233**, 532.
36. Assoian,R.K., Komoriya,A., Meyers,C.A., Miller,D.M. and Sporn,M.B. (1983) *J. Biol. Chem.,* **258**, 7155.
37. Massague,J. (1985) *Prog. Med. Virol.,* **32**, 142.
38. Massague,J. (1984) *J. Biol. Chem.,* **259**, 9756.
39. Frolik,C.A., Dart,L.L., Meyers,C.A., Smith,D.M. and Sporn,M.B. (1983) *Proc. Natl. Acad. Sci. USA,* **80**, 3676.
40. Heldin,C.H., Wasteson,A. and Westermark,B. (1985) *Molecular and Cellular Endocrinology,* **39**, 169.
41. Raines,E.W. and Ross,R. (1982) *J. Biol. Chem.,* **267**, 5154.
42. Heldin,C.H., Westermark,B. and Wasteson,A. (1981) *Biochem. J.,* **193**, 907.
43. Sugarman,B.J., Aggarwal,B.B., Hass,P.E., Figari,I.S., Palladino,M.A. and Shepard,M.H. (1985) *Science,* **30**, 943.
44. Old,L.J. (1985) *Science,* **230**, 630.
45. Wong,G.H. and Goeddel,D.W. (1986) *Nature,* **323**, 819.
46. Aggarwal,B.B., Kohr,W.J., Hass,P.E., Moffat,S.A., Henzel,W.J., Bringman,T.S., Nedwin,G.E., Goeddel,D.V. and Harkins,R.N. (1985) *J. Biol. Chem.,* **260**, 2345.
47. Shirai,T., Yamagushi,H., Hirataka,I., Todd,C.W. and Wallace,B. (1985) *Nature,* **313**, 803.
48. Lobb,R.R., Harper,J.W. and Fett,J.W. (1986) *Anal. Biochem.,* **154**, 1.
49. Lobb,R.R., Alderman,E.M. and Fett,J.W. (1985) *Biochemistry,* **24**, 4969.
50. Gospodarowicz,D. and Cheng,J. (1986) *J. Cellular Physiol.,* **128**, 475.
51. George-Nasclmento,C., Gyenes,A., Halloran,S.M., Merryweather,J., Valenzuela,P., Steimer,K.S., Masiarz,F.R. and Randolph,A. (1988) *Biochemistry,* **27**, 797.
52. Esch,F., Baird,A., Ling,N., Veno,N., Hill,F., Denoroy,L., Kleepper,R., Gospodarowicz,D., Böhlen,P. and Guillemic,R. (1985) *Proc. Natl. Acad. Sci. USA,* **82**, 6507.
53. Duance,V.C. and Bailey,A.J. (1981) Biosynthesis and degradation of collagen. In *Handbook of Inflammation.* Glynn,L.E., Houck,J.C. and Weissman,G. (eds), **3**, 51.
54. Cunningham,L.W. and Frediksen,D.W. (eds) (1982) *Methods in Enzymology,* Academic Press, New York, Vol. 82, Part A.
55. Light,N.D. (1985) In *Methods in Skin Research.* Skerrow,D. and Skerrow,C.J. (eds), p. 559.
56. Bornstein,P. and Sage,H. (1980) *Ann. Rev. Biochem.,* **49**, 957.
57. Byers,P.H., McKenney,K.H., Lichtenstein,J.R. and Martin,G.K. (1974) *Biochemistry,* **13**, 5243.
58. Sage,H. and Bornstein,P. (1982) In *Methods in Enzymology.* Cunningham,L.W. and Frederiksen,D.W. (eds) Academic Press, New York, Vol. 82, p. 96.
59. Herbage,D., Bouillet,J. and Bernengo,J.-C. (1977) *Biochem. J.,* **161**, 303.
60. Timpl,R., Martin,G.R., Bruckner,P., Wick,G. and Wiedmann,H. (1978) *Eur. J. Biochem.,* **84**, 43.
61. Duance,V.C., Wotton,S.F., Voyle,C.A. and Bailey,A.J. (1984) *Biochem. J.,* **221**, 885.
62. Bentz,N., Bachinger,H., Glanville,R. and Kuhn,K. (1978) *Eur. J. Biochem.,* **92**, 563.
63. Jander,R., Rauterberg,J., Voss,B., von Bassewitz,D.B. (1981) *Eur. J. Biochem.,* **114**, 17.
64. Odermatt,E., Risteli,J., van Delden,V. and Timpl,R. (1983) *Biochem. J.,* **211**, 295.
65. Fleischmajer,R., Olsen,B. and Kuhn,K. (eds) (1985) *Ann. N.Y. Acad. Sci.,* **460**.
66. Light,N.D. and Bailey,A.J. (1985) In *Methods in Enzymology.* Cunningham,L.W. and Frederiksen,D.W. (eds), Academic Press, New York, Vol. 82, 360.
67. Piez,K.A. (1968) *Anal. Biochem.,* **26**, 305.
68. Bailey,A.J., Sims,T.J., Duance,V.C. and Light,N.D. (1979) *FEBS Lett.,* **99**, 361.
69. Burgeson,R., El Adli,F., Kaitila,I. and Hollister,D. (1976) *Proc. Natl. Acad. Sci. USA,* **73**, 2579.

70. Shimokomaki,M., Duance,V.C. and Bailey,A.J. (1980) *FEBS Lett.*, **121**, 51.
71. Shimokomaki,M., Duance,V.C. and Bailey,A.J. (1981) *Biosci. Rep.*, **1**, 561.
72. Okada,Y., Nagase,H. and Harris,E.D. (1986) *J. Biol. Chem.*, **261**, 14245.
73. Cawston,T.E. and Tyler,J.A. (1979) *Biochem. J.*, **183**, 647.
74. Cawston,T.E. and Murphy,G. (1981) In *Methods in Enzymology*. Colowick,S.P. and Lorand,L. (eds), Academic Press, New York, Vol. 80, p. 711.
75. Galloway,W.A., Murphy,G., Sandy,J.D., Gavrilovic,J., Cawston,T.E. and Reynolds,J.J. (1983) *Biochem. J.*, **209**, 741.
76. Hibbs,M.S., Hasty,K.A., Seyer,J.M., Kang,A.H. and Mainardi,C.L. (1985) *J. Biol. Chem.*, **260**, 2493.
77. Seltzer,J.L., Eschbach,M.L. and Eisen,A.Z. (1985) *J. Chromatogr.*, **326**, 147.
78. Wilhelm,S.M., Collier,I.E., Kronberger,A., Eisen,A.Z., Marmer,B.L., Grant,G.A., Bauer,E.A. and Goldberg,G.I. (1987) *Proc. Natl. Acad. Sci. USA*, **84**, 6725.
79. Birkedal-Hansen,H. (1987) In *Methods in Enzymology*. Academic Press, New York, Vol. 144, p. 140.
80. Collier,I.E., Wilhelm,S.M., Eisen,A.Z., Marmer,B.L., Grant,G.A., Seltzer,J.L., Kronberger,A., He,C., Bauer,E.A. and Goldberg,G.I. (1988) *J. Biol. Chem.*, **263**, 6579.
81. Moore,W.M., Spilburg,C.A., Hirsch,S.K., Evans,C.L., Wester,W.N. and Martin,R.A. (1986) *Biochemistry*, **25**, 5189.
82. Murphy,G., Cockett,M.I., Stephens,P.E., Smith,B.J. and Docherty,A.J.P. (1987) *Biochem. J.*, **248**, 265.
83. Weigele,M., DeBernado,S.L., Tengi,J.P. and Leimgruber,W. (1972) *J. Am. Chem. Soc.*, **94**, 5927.
84. Laemmli,U.K. and Favre,M. (1973) *J. Mol. Biol.*, **80**, 575.
85. Merril,C.R., Goldman,D., Sedman,S.A. and Ebert,M.H. (1981) *Science*, **211**, 1437.
86. Towbin,H., Staehelin,T. and Gordon,J. (19   ) *Proc. Natl. Acad. Sci. USA*, **76**, 4350.
87. Bradford,M.M. (1976) *Anal. Biochem.*, **72**, 248.
88. Cleveland,W.L., Wood,I. and Erlanger,B.F. (1983) *J. Immunol. Methods*, **56**, 221.
89. Dixon,M. and Webb,E.C. (1961) *Adv. Prot. Chem.*, **16**, 197.
90. Carter,R.J. and Boyd,N.D. (1979) *J. Immunol. Methods*, **26**, 213.
91. Phillips,A.P., Martin,K.L. and Horten,W.H. (1984) *J. Immunol. Methods*, **74**, 385.
92. Ey,P.L., Prowse,S.J. and Jenkin,C.R. (1978) *Immunochemistry*, **15**, 429.
93. Rousseaux,J., Picque,M.T., Baxin,H. and Bisette,G. (1981) *Mol. Immunol.*, **18**, 639.
94. Stephenson,J.R., Lee,J.M. and Wilton-Smith,P.D. (1984) *Anal. Biochem.*, **142**, 189.
95. Akerstrom,B. and Bjorck,L. (1986) *J. Biol. Chem.*, **261**, 10240.
96. O'Sullivan,M.J., Gnemimi,E., Chietegatti,G., Morris,D., Simmonds,A.D., Simmonds,M., Bridges,J.W. and Marks,V. (1979) *J. Immunol. Methods*, **30**, 127.
97. Fahey,J.L. and Terry,E.W. (1978) In *Handbook of Experimental Immunology*. Weir,D.M. (ed.), Blackwell, London, p. 81.
98. Gorbunoff,M.J. and Timasheff,S.W. (1984) *Anal. Biochem.*, **136**, 440.
99. Stanker,L.H., Vanderlaan,M. and Juarez-Salinas,H. (1985) *J. Immunol. Methods*, **76**, 157.
100. Goheen,S.C. and Chow,T. (1986) *J. Chromatog.*, **359**, 257.
101. Nissonoff,A., Wissler,F.C., Lipman,L.M. and Woernley,D.L. (1960) *Arch. Biochem. Biophys.*, **89**, 230.
102. Lamoyi,E. (1986) In *Methods in Enzymology*. Langone,J.J. and Vunakis,H.V. (eds), Academic Press, New York, Vol. 121, p. 652.
103. Parham,P., Androlewicz,M.J., Brodsky,F.M., Holmes,N.J. and Ways,J.P. (1982) *J. Immunol. Methods*, **53**, 133.
104. Young,W.W.Jr., Tamura,Y., Wolock,D.M. and Fox,J.W. (1984) *J. Immunology*, **133**, 3163.

APPENDIX I
Suppliers

APV, P.O. Box 4, Manor Royal, Crawley, West Sussex, RH10 2QB, UK.

Aldrich, The Old Brickyard, New Road, Gillingham, Dorset, SP8 4JL, UK. Tel. (07476) 2211.

Amersham, Lincoln Place, Green End, Aylesbury, Bucks., HP20 2TP, UK. Tel. (0296) 395222.

Amicon, Upper Mill, Stonehouse, Gloucester, GL10 2BJ, UK. Tel. (045382) 5181.

Anachem, 20 Charles Street, Luton, Beds., LU2 0EB, UK. Tel. (0582) 456666.

Anderman & Co. Ltd, 145 London Road, Kingston upon Thames, Surrey, KT2 6NH, UK. Tel. (01) 541 0035.

Artisan Metal Products, Waltham, MA, USA.

Atlas Bioscan Ltd, Osborne House, Stockbridge Road, Chichester, PO19 2DU, UK. Tel. (0243) 773141.

BCL, Bell Lane, Lewes, East Sussex, UK. Tel. (0273) 480444.

BDH, Broom Road, Parkstone, Poole, Dorset, UK. Tel. (0202) 745520.

Beckman, Progress Road, Sands Ind. Estate, High Wycombe, Bucks., HP12 4JL, UK.

Biocatalysts Ltd, Tredforest Industrial Estate, Pontypridd, Wales, UK.

BioInvent International, S-223, 70 Lund, Sweden. Tel. (046) 46 16 85 50.

BioProbe International Inc., 2842 Walnut Avenue, Tustin, CA 92680, USA. Tel. (714) 544 4035.

BioRad Laboratories Ltd, Caxton Way, Holywell Ind. Estate, Watford, Herts., WD1 8RP, UK. Tel. (0923) 240 322.

BRL, PO Box 145, Science Park, Cambridge, CB4 4BE, UK.

Brownlea Labs. Inc., 2045 Martin Ave., Santa Clara, CA 95050, USA. Tel. (408) 727 1346.

Calbiochem, see Nova Biochem, UK.

Cambio, 34 Millington Road, Newnham, Cambridge CB3 9HP, UK. Tel. (0223) 66500.

Cecil Instruments Ltd, Milton Technical Centre, Milton, Cambridge, CB4 4AZ, UK. Tel. (0223) 420821.

Christison Scientific Equipment Ltd, Albany Road, S.H.House, East Gateshead Industrial Estate, Gateshead, NE8 3AT, UK. Tel. (0632) 77461.

Collaborative Research, PO Box 370068, Boston, MA 02241, USA.

Cooper Biomedical & ICN Immunobiologicals, Free Press House, Castle Street, High Wycombe, Bucks., HA3 6RN, UK. Tel. (0494) 443826.

Cuthbert Andrews Ltd, Watford, UK.

Dakopatts Ltd, 22 The Arcade, The Octagon, High Wycombe, Bucks., HP11 2HT, UK. Tel. (0494) 452016.

DDS Membrane Filtration, 1600 County Road, F.Hudson, WI 54016, USA. Tel. (715) 386 9371.

Desaga, Springfield Mill, Sandling Road, Maidstone, Kent, ME14 2LE, UK.

Dominick Hunter Ltd, Durham Road, Birtley, DH3 2SF, UK. Tel. (091) 4105121, see also HPLC Technology Ltd.

Drew Scientific Ltd, 12 Barley Mow Passage, London, W4 4PH, UK. Tel. (01) 995 9382.

DuPont, Wedgewood Way, Stevenage, Herts., SG1 4QN, UK. Tel. (0438) 734680.

Eastman Kodak, LRPD, Acorn Field Road, Liverpool, L33 7ZX, UK.

Edwards, Manor Royal, Crawley, Sussex.

EDT Analytical, 14 Trading Estate Road, London, NW10 7LU, UK. Tel. (01) 961 1477.

Electro-Nucleonics International Ltd, Adriann van Bergerstraat, 202−208, 4811 SW Breda, Netherlands. Tel. (7622) 2033.

Eppendorf, see Anderman.

Filtron Technology Corp., 500 Main St., PO Box 119, Clinton, MA 01510, USA. Tel. (508) 3688582.

Fisher Scientific, 711 Forbes Ave., Pittsburg, PA, USA. Tel. (412) 562 8300.

Fisons Instruments, Sussex Manor Park, Gatwick Road, Crawley, RH10 2QQ, UK. Tel. (0293) 561222.

Flow Laboratories, Woodcock Hill, Harefield Road, Rickmansworth, Herts., WD3 1PQ, UK. Tel. (0923) 774666.

Fluka, Peakdale Road, Glossop, Derby, SK13 9XE, UK.

Gallenkamp, Belton Road, Loughborough, Leics., LE11 0TR, UK.

Gelman Sciences Ltd, 10 Harrowdean Road, Brackmills, Northampton, NN4 0EZ, Tel. (0604) 765141.

Gibco UK Ltd, PO Box 35, Trident House, Renfrew Road, Paisley, PA3 4EF, Scotland, UK. Tel. (041) 889 6100.

Glen Creston Ltd, 16 Dalston Gardens, Stanmore, Middlesex, HA7 1DA, UK. Tel. (01) 226 0123.

Heraeus Equipment Ltd, Unit 9, Wates Way, Brentwood, Essex, CM15 9TB, UK. Tel. (0509) 237371.

Hoeffer Scientific Instruments, Unit 12, Croft Road Workshops, Croft Road, Newcastle under Lyme, ST5 0TH, UK. Tel. (0782) 617317.

Hoeffer (Biotech), 183A Camford Way, Luton, Beds. LU3 3ANS, UK

IBF, see Life Science Laboratories Ltd.

ICN Biomedicals Ltd, Free Press House, Castle St., High Wycombe, Bucks., HP13 6RN, UK. Tel. (0494) 443826.

Janssen, Grove, Wantage, Oxon. OX12 0DQ, UK. See also ICN.

Jencons, Cherrycourt Way Ind. Estate, Leighton Buzzard, Beds., LU7 8UA, UK.

Joyce Loebel, Marquisway, Team Valley, Gateshead, NE11 0QW, UK. Tel. (091) 482 2111.

J.T.Baker, PO Box 9, Hayes Gate House, 27 Uxbridge Road, Hayes, Middlesex, UB4 JD, UK. Tel. (01) 569 1191.

Kabivitrum, Kabi Vitrum House, Riverside Way, Uxbridge, UB8 2YF, UK.

Life Science Laboratories, Sedgewick Road, Luton, LU4 9DT, UK. Tel. (0582) 597676.

LKB, see Pharmacia−LKB.

Merck, see BDH.

Microfluidics Corp., 44 Mechanic St., Newton, MA 02164, USA. Tel. (617) 969 5452.

Microgen Inc., 23152 Verdugo Dr., Laguna Hills, CA 92653, USA. Tel. (714) 581 3880.

Miles Ltd, Stoke Court, Stoke Poges, SL2 4LY, UK. Tel. (02814) 5151.

Millipore Waters, The Boulevard, Ascot Road, Croxley Green, Watford WD1 8YW, UK. Tel. (0923) 816375.

MSE Scientific Instruments, Sussex Manor Park, Crawley, Sussex, RH10 2QQ, UK. Tel. (0293) 31100.

New Brunswick Scientific (UK) Ltd, 6 Colonial Way, Watford, WD2 4PT, UK. Tel. (0923) 223293.

Northern Media Supply, Sainsburyway, Hessle, N. Humberside, HU13 9NX, UK. Tel. (0482) 572436.

Nova Biochem, 3 Heathcote Buildings, Highfields Science Park, University Blvd., Nottingham, NG7 2QJ, Tel. (0602) 430951.

Novo Ltd, 28 Thomas Avenue, Windsor, Berks., SL1 1QP, UK.

Nucleopore Corp., 7035 Commerce Cr., Pleasanton, CA 94566, USA. Tel. (415) 463 2530.

PCI, One Fairfield Crescent, West Caldwell, NJ 07006, USA. Tel. (201) 575 7052.

Perkin-Elmer Ltd, Post Office Lane, Beaconsfield, Bucks., HP9 1QA, UK. Tel. (0494) 676161.

Perstorp Biolytica, S-223 70 Lund, Sweden. Tel. (046) 46 16 87 80. See also EDT Analytical.

Pfeiffer and Langen Dormagen, Frankenstrasse 25, D-4047 Dormagen, FRG. Tel. (02106) 52-1.

Pharmacia − LKB, Pharmacia House, Midsummer Boulevard, Milton Keynes, MK9 3HP, UK. Tel. (0908) 661101.

Phase-Separations Ltd, Deeside Industrial Park, Queensferry, Clwyd, CH5 2NU, UK. Tel. (0244) 816444.

Phillips Analytical, Building HKF, NL-5600 MD Eindhoven, The Netherlands. Tel. (040) 785213.

Pierce Europe BV, PO Box 1512, NL 3260, The Netherlands. See also Life Science Laboratories.

Rohm Pharma GmbH, Westerstadt, PO Box 4347, D-6100, Darmstadt 1, FRG. Tel. (06151) 877-0.

Romicon (Rohm & Haas Ltd), Lennig House, 2 Masons Avenue, Croydon, CR9 3NB, UK.

Russell pH Ltd, Station Road Auchtermuchty, Fife KY14 7DP, Scotland, UK. Tel. (03372) 8871.

Sarstedt, 68 Boston Road, Leicester, LE4 1AW, UK.

Sartorius, 18 Avenue Road, Belmont, Sutton, Surrey, UK.

Schleicher and Schuell GmbH, P.O. Box 246, D-3352, Einbeck, FRG.

Scientific Supplies, 618 Western Avenue, Park Royal, London, W3 OTE, UK.

Serva, see Cambridge Bioscience.

Shandon Southern Products Ltd, Chadrich Road, Astmoor, Runcorn, Cheshire, WA7 1PR, UK. Tel. (09285) 66611.

Shimadzu Scientific Instruments, 7102 Riverwood Road, Columbia, MD 21046, USA. Tel. (301) 381 1227.

Sigma, Fancy Road, Poole, Dorset, BH17 7NH, UK. Tel. (0202) 733114.

Sorvall-DuPont, see DuPont.

Sterogene Biochemicals, 136 E. Santa Clara St., Arcadia, CA 91006, USA. Tel. (818) 446 3773.

Sturge Ltd, Denison Road, Sellay, North Yorkshire, YO8 8EF, UK.

Synchrom Inc., see Anachem.

Toyo Soda, see Anachem.

Ulvac (Chemlab), Hornminster House, 129 Upminster Road, Hornchurch, Essex, KM11 3XJ, UK.

Uniscience Ltd, 12−14 St. Annes Crescent, London SW18 2LS, UK.

Union Carbide Corp., Old Ridgebury Road, Danbury, CT 06817, USA. Tel. (203) 7945300.

Varian Associates Ltd, 28 Manor Road, Walton on Thames, Surrey, KT12 2QF, UK. Tel. (09322) 43741.

Wako, see Cambridge Bioscience.

Watson Marlow, Smith and Nephew Pharmaceuticals Ltd, Falmouth, Cornwall, TR11 4RW, UK.

Wellcome Diagnostics, Temple Hill, Dartford, Kent, DA1 5AH, UK.

Whatman Lab Sales Ltd, Unit 1, Coldred Road, Parkwood, Maidstone, Kent, ME15 9XN, UK. Tel. (0622) 674821.

INDEX